Web Development with Julia and Genie

A hands-on guide to high-performance server-side web development with the Julia programming language

Ivo Balbaert

Adrian Salceanu

BIRMINGHAM—MUMBAI

Web Development with Julia and Genie

Group Product Manager: Pavan Ramchandani
Publishing Product Manager: Himani Dewan
Senior Editor: Divya Anne Selvaraj
Technical Editor: Saurabh Kadave
Copy Editor: Safis Editing
Project Coordinator: Manthan Patel
Proofreader: Safis Editing
Indexer: Subalakshmi Govindhan
Production Designer: Joshua Misquitta
Marketing Coordinator: Anamika Singh

First published: November 2022
Production reference: 1231122

Published by Packt Publishing Ltd.
Livery Place
35 Livery Street
Birmingham
B3 2PB, UK.

ISBN 978-1-80181-113-2

www.packt.com

Foreword

Julia has long been viewed as a tool that was merely for niche scientific computing, and if you wanted to do something beyond crunching numbers, you were out of luck. *Ivo Balbaert's* and *Adrian Salceanu's* work had to dispel this myth in *Web Development with Julia and Genie*. I have had the chance to interact with Adrian over the last few years in my role supporting the Julia community, and it has blown me away to see his conviction that Julia developers (and more broadly those in the computing ecosystem) should have access to world-class web development tools in the same language as their computational tools.

Adrian's deep conviction has given birth to the *Genie Framework*, a production-ready ecosystem to build web apps natively in Julia. But Adrian's belief in this space goes beyond his open source contributions. He has also made a big bet with *Genie Cloud*, the company he co-founded, which is trying to bring this unique mix of computational tools and web expertise to developers through a low code interface.

There are no people better suited to teach the information in this book than this book's authors. By learning directly from the creator of the framework, you will be able to learn the philosophy of the ecosystem which extends well beyond simple usage examples that you would find in traditional documentation. Adrian also has extensive experience in the Julia ecosystem, which he garnered over many years of maintaining and contributing to various packages and Julia projects.

I am absolutely thrilled to see the *Genie Framework's* usage growing and continuing to challenge the narrative that Julia is only for scientific computing. I expect that readers of this book will come away with a tremendous amount of respect for just how impressive the *Genie Framework* is, and all the hard work it took from Adrian and other contributors. I am so glad to have Adrian and Ivo as part of the Julia community and look forward to seeing the impact of this book and the *Genie Framework* in the years to come.

Logan Kilpatrick

Julia Project, Lead Developer Community Advocate

Contributors

About the authors

Ivo Balbaert is a lecturer in web programming and databases at CVO Antwerpen, a community college in Belgium. He received a PhD in applied physics from the University of Antwerp in 1986. He worked for 20 years in the software industry as a developer and consultant in several companies, and for 10 years as project manager at the University Hospital of Antwerp. From 2000 onwards, he switched to partly teaching and partly developing software (at KHM Mechelen, CVO Antwerpen). He also wrote an introductory book in Dutch about developing in Ruby and Rails, *Programmeren met Ruby en Rails*, published by Van Duuren Media. In 2012, he authored a book on the Go programming language, *The Way to Go*, published by iUniverse. He has written a number of introductory books for new programming languages, notably Dart, Julia, Rust, and Red, all published by Packt.

I want to thank the reviewers for helping to increase the value and accuracy of the book, and a special thanks to our senior editor, Divya Anne Selvaraj, for her relentless devotion to improving the text.

Adrian Salceanu is the creator and lead maintainer of *Genie Framework*. He has over two decades of professional work experience as a web developer and software architect, leading agile teams in developing, scaling, and maintaining business critical, data-intensive web applications. Currently, he is the technical founder and CEO of *Genie Cloud*, a no-code app development platform built with Genie. Adrian is the author of *Julia Programming Projects* (published by Packt in 2018) and an enthusiastic *JuliaLang* open-source contributor. He has two master's degrees, one in Computing, and another in Advanced Computer Science.

About the reviewers

Abhimanyu Aryan is a Julia developer at Genie Framework OSS. Previously, he worked as an ML engineer for an aerospace company for defense services and taught full-stack JavaScript at India's leading boot camp. He started his career as a Unity developer, working on various augmented reality and virtual reality use cases for Fortune 500 companies.

He also contributed a chapter on Android in *Economics of Information Technology: Emerging Trends and Prospects*, by Dr. Sunil Phougat and Dr. Pushp Deep Dagar, Savera Publishing House, 2015.

Mattia Nicolò Careddu is a software and data engineer from Milan, Italy, and a Genie contributor. Currently, he is the lead AI engineer and software engineer at nCore, an HR scaleup based in Milan, while also pursuing a bachelor's degree in statistics and big data. He also founded PlayLife Events, a tech start-up in the event world, based in Italy. He is passionate about technology, AI, and start-ups.

Table of Contents

3

Applying Julia in Various Use Cases on the Web   53

Part 2: Using the Genie Rapid Web Development Framework

4

Building an MVC ToDo App   77

5

Adding a REST API   115

6

Deploying Genie Apps in Production   139

7

Adding Authentication to Our App   171

8

Developing Interactive Data Dashboards with Genie   195

Preface

This book provides an overview of web development in Julia and teaches you how to build MVC applications with a REST API and an interactive data dashboard using the Genie web framework.

Who this book is for

The intended audience has web development experience, preferably some experience with Julia as well, and wants to blend these to become able to write web applications with Julia.

What this book covers

Chapter 1, *Julia Programming Overview*, reviews the Julia programming language in order to prepare you for web development with Julia. We'll work with Julia in the REPL and with the Visual Studio Code editor, which is how we'll build web apps in the rest of the book. Then, we will look at the types, flow controls, functions, and methods that you'll need in any Julia app.

Chapter 2, *Using Julia Standard Web Packages*, delves into what Julia's standard library and JuliaWeb have to offer for building web apps. First, we will look at working with the Sockets package and building two versions of a TCP echo server. Then, we will examine how to work with the WebSockets package, followed up with a thorough discussion of the HTTP package. Next, we'll build several web server examples and see how to route incoming requests. Finally, it will all come together when working with JSON data using a REST web server.

Chapter 3, *Applying Julia in Various Use Cases on the Web*, examines what the Julia community has built over the past 5 years in the way of tools to build web-based applications. To make a static website, Franklin is available to you as a valuable choice. If you need to build a cross-platform Electron-based GUI app, you can combine WebIO, Blink, and Interact to accomplish this. If your next project is to build a web dashboard, look no further than Dash for Julia. We will also discuss the architecture of microservices, the current state of WASM in the Julia ecosystem, and which cloud platforms are readily available for Julia. Finally, you will get an overview of all the mature web frameworks in Julia, giving you a better idea of which to apply when and gain a quick start to working with any of them.

Chapter 4, *Building an MVC ToDo App*, guides us through creating a complete application in the Genie Julia web framework, which you can take as a blueprint for starting your own project. You'll learn how to generate a new MVC project and its model, how to create views to display the data, and how to route the browser requests through a controller function to its intended view. You'll also see how you can add data validation functions and set up automatic testing for your app.

Chapter 5, Adding a REST API, enriches our to-do application with a REST API. You'll learn how to code the CRUD functions for this API. The REST API exposes your data to the world – that's why testing and documenting are especially important. You'll learn how to write these specific tests, and how to document the API using Swagger.

Chapter 6, Deploying Genie Apps in Production, discusses the different possibilities for deploying Genie apps. First, we'll see how to customize a production environment and then learn how to package our app in a Docker container, how to set up a repository on GitHub with the automatic execution of tests (as in, CI), and how to deploy our app in a Docker container using Git. We'll also learn how to improve the app's startup by using `PackageCompiler.jl`. Finally, we will explore an alternative method of deployment using Heroku.

Chapter 7, Adding Authentication to Our App, takes us through adding an authentication layer to our app, so that only registered users will be able to use it, and a user will only see their own to-do items. Also, we will expand our tests for this new functionality.

Chapter 8, Creating an Interactive Data Dashboard, explores using the Genie Builder tool to build a dashboard to-do application. We'll learn how to generate mock data, create and visualize a time series, and make various types of charts and filters.

To get the most out of this book

You will need a version of *Julia* installed on your computer – preferably *Julia 1.8*. You can install it from `https://julialang.org/downloads/`.

Software/hardware covered in the book	Operating system requirements
Julia 1.8	Windows, macOS, or Linux
Visual Studio Code	
Genie 5	

All code examples have been tested using *Julia 1.8* on *Windows OS* and *Linux*. *Visual Studio Code* can be downloaded from `https://code.visualstudio.com/`.

For *Part 2* of this book (*Chapters 4-8*), make sure you use the most up-to-date version of the *Genie web development framework* (version 5, at the time of writing this book).

If you are using the digital version of this book, we advise you to type the code yourself or access the code from the book's GitHub repository (a link is available in the next section). Doing so will help you avoid any potential errors related to the copying and pasting of code.

Download the example code files

You can download the example code files for this book from GitHub at `https://github.com/PacktPublishing/Web-Development-with-Julia-and-Genie`. If there's an update to the code, it will be updated in the GitHub repository.

We also have other code bundles from our rich catalog of books and videos available at `https://github.com/PacktPublishing/`. Check them out!

Download the color images

We also provide a PDF file that has color images of the screenshots and diagrams used in this book. You can download it here: `https://packt.link/PymYa`.

Conventions used

There are a number of text conventions used throughout this book.

`Code in text`: Indicates code words in text, database table names, folder names, filenames, file extensions, pathnames, dummy URLs, user input, and Twitter handles. Here is an example: "If the file has no header line, specify the `header=false` keyword argument."

A block of code is set as follows:

```
if todo2.priority > todo1.priority
    println("Better do todo2 first")
else
    println("Better do todo1 first")
end
```

When we wish to draw your attention to a particular part of a code block, the relevant lines or items are set in bold:

```
<h4 class="container">
Sorry, no results were found for "$(params(:search_movies))"
</h4>
```

Any command-line input or output is written as follows:

```
user@DESKTOP$ nc localhost 8080
Hello Server, can you hear me?          # client message
```

Bold: Indicates a new term, an important word, or words that you see onscreen. For instance, words in menus or dialog boxes appear in **bold**. Here is an example: "Search in the **Extensions** tab for Julia and install it."

> **Tips or important notes**
> Appear like this.

Get in touch

Feedback from our readers is always welcome.

General feedback: If you have questions about any aspect of this book, email us at `customercare@ packtpub.com` and mention the book title in the subject of your message.

Errata: Although we have taken every care to ensure the accuracy of our content, mistakes do happen. If you have found a mistake in this book, we would be grateful if you would report this to us. Please visit `www.packtpub.com/support/errata` and fill in the form.

Piracy: If you come across any illegal copies of our works in any form on the internet, we would be grateful if you would provide us with the location address or website name. Please contact us at `copyright@packt.com` with a link to the material.

If you are interested in becoming an author: If there is a topic that you have expertise in and you are interested in either writing or contributing to a book, please visit `authors.packtpub.com`.

Share your thoughts

Once you've read *Web Development with Julia and Genie*, we'd love to hear your thoughts! Scan the QR code below to go straight to the Amazon review page for this book and share your feedback.

`https://packt.link/r/1-801-81113-X`

Your review is important to us and the tech community and will help us make sure we're delivering excellent quality content.

Download a free PDF copy of this book

Thanks for purchasing this book!

Do you like to read on the go but are unable to carry your print books everywhere?

Is your eBook purchase not compatible with the device of your choice?

Don't worry, now with every Packt book you get a DRM-free PDF version of that book at no cost.

Read anywhere, any place, on any device. Search, copy, and paste code from your favorite technical books directly into your application.

The perks don't stop there, you can get exclusive access to discounts, newsletters, and great free content in your inbox daily!

Follow these simple steps to get the benefits:

1. Scan the QR code or visit the link below:

https://packt.link/free-ebook/9781801811132

2. Submit your proof of purchase
3. That's it! We'll send your free PDF and other benefits to your email directly

Part 1:
Developing
Web Apps with Julia

By the end of this part, you will understand the core components of the Julia programming language that serve as a basis for web development.

Need help with basic concepts of Julia? Or looking to learn more about its importance? The first three chapters are all about the basic principles of Julia, helping you to understand why Julia is a good fit for web development.

By the end of this part of the book, you will be able to use the foundational Julia packages such as Sockets, JSON, JSON3, HTTP, and WebSockets, and combine them in writing a REST services endpoint to-do app.

Moreover, you'll explore building a static website with Franklin; creating a dashboard with Dash; how Julia can interact with JavaScript, microservices, and cloud services; and gain insights regarding choosing the right Julia framework for a project.

This part of the book comprises the following chapters:

- *Chapter 1, Julia Programming Overview*
- *Chapter 2, Using Julia Standard Web Packages*
- *Chapter 3, Applying Julia in Various Use Cases on the Web*

1

Julia Programming Overview

Julia is a high-performance, open source computing language, mostly applied to data analysis, machine learning, and other scientific and technical computing applications.

The language combines the ease of use of **Python** or **R** with the speed of **C** and eliminates the need for using two languages to develop data intensive applications. It is as readable and high-level as Python and because of its type inference and optional typing, behaves as a dynamic language. It is also as fast as C, but much more readable. As a new programming language, Julia borrowed some of the best features from other modern languages. For example, like Ruby, it doesn't use semicolons or curly braces for delimiting code; instead, it uses a more Pascal-like syntax, with `end` to indicate where a code structure stops.

Julia is not a classic object-oriented language like *Java*; instead, it is more *function-oriented*, but it also has a struct data type like C. Functions that act on and transform data are the basic building blocks. The language also has built-in parallel computing capabilities and can scale up very easily.

Julia also provides an extensive standard library from the start. The language's usage and popularity are steadily rising; it has been downloaded by users from more than 10,000 companies and is used at over 1,500 universities worldwide (`https://juliacomputing.com/media/2022/02/julia-turns-ten-years-old/`).

This chapter will touch on the main Julia concepts we will need in web development including types, flow control, functions, packages, and modules. We will introduce some examples relating to the ToDo app project theme for *Part 2* of this book.

We'll also show code snippets from the Genie framework that are used in *Part 2*. We wrap up with a section on how Julia works internally, which makes us better understand Julia's efficacy in web development. By the end of this chapter, your Julia knowledge will be refreshed, and you'll be much better prepared to grasp the rest of the book.

In this chapter, we will cover the following topics:

- Working with Julia
- Types, flow controls, and functions in Julia
- Useful techniques in Julia web development

- Using Julia modules and packages
- How Julia works
- Why Julia is a good fit for web development

Technical requirements

To follow through with all the exercises in this chapter and the rest of the book, you will need the following:

- Access to the command-line terminal on your OS.

- The latest current stable release of Julia installed. Install Julia on your computer by downloading the latest current stable release (*v 1.8* or higher) from `https://julialang.org/downloads/` or through the cross-platform *juliaup* tool (`https://github.com/JuliaLang/juliaup`).

- **VS Code**: Download it from `https://code.visualstudio.com/`, preferably *v1.66* or higher.

All the code examples in this book have been run on a Windows machine. You may find subtle differences in output if you are using Linux/macOS. Where necessary, command variations for both machines have been specified.

The complete source code for this chapter can be found at `https://github.com/PacktPublishing/Web-Development-with-Julia-and-Genie/tree/main/Chapter1`.

Working with Julia

In this section, we will set up a standard Julia development environment, and learn how to work with Julia scripts as well as the **Read–Eval–Print Loop** (**REPL**). The REPL allows you to work with Julia in an interactive way, trying out expressions, function calls, and even executing whole programs.

You're good to go when typing in `julia` at the terminal starts up the REPL:

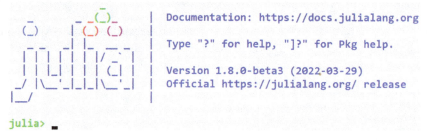

Figure 1.1 – The Julia REPL

Using the REPL to use Julia interactively

Try typing in 256^2, giving the result 65536, or rand(), which gives you a random number between 0 and 1, for example, 0.02925477322848513. We'll use the REPL extensively throughout the book, and the **Genie web framework** discussed in *Part 2* allows you to build your entire web app from the REPL.

Some useful REPL keyboard shortcuts you will use often include the following:

- *Ctrl + D*: To exit the REPL
- *Ctrl + L*: To clear the screen
- *Ctrl + C*: To get a new prompt
- Up and down arrows: To reuse recent commands

To see which folder you are in, type pwd(). To change the current folder, execute cd("path/to/folder").

A feature we'll use a lot in Genie is creating new files from within the REPL with the touch command. For example, to create an empty testset.jl file in an existing folder structure, testing/phase1, enter the following:

```
julia> touch(joinpath("testing", "phase1", "testset.jl"))
```

The REPL returns testing\\phase1\\testset.jl on Windows and testing/phase1/testset.jl on *nix* systems.

The joinpath function constructs the directory path starting in the current folder, and touch creates the file.

Using the package mode to jump-start a project

The Julia ecosystem encompasses thousands of libraries, called packages (see https://julialang.org/packages/), for which Julia has a built-in package manager, *Pkg*.

The REPL has a special mode for working with packages, which is started by typing] at julia> prompt, which brings you to package mode: (@v1.8) pkg>.

Some useful commands in this mode to type in after the pkg> prompt are as follows:

- st or status: Gets a list of all the packages installed in your environment.
- add PackageName: Adds a new package (you can add several packages separated by , if needed).
- up or update: Updates all your packages.

- `up or update PackageName`: Updates a specific package.
- `activate .`: Activates the current project environment (see the *Packages and projects* section under *Using Julia modules and packages*). `rm or remove PackageName`: Removes a specific package.
- `?`: Lists all available commands.
- The *backspace* key: Exits the `pkg>` mode.

In the next section, *Parsing a CSV file*, we will work with a **comma-separated values** (**CSV**) file of to-do items. The CSV package can be imported and set up to be used in your Julia REPL by typing the following:

```
julia> using Pkg
julia> Pkg.add("CSV")
julia> using CSV
```

The last line of the preceding command brings the definitions of the CSV package into scope.

Alternatively, from the package mode, use the following:

```
]:
(@v1.8) pkg> add CSV
```

The preceding command installs all packages CSV depends on and then precompiles them. This way, the project gets a jump-start, because the **just-in-time** (**JIT**) **compiler** doesn't have to do this work anymore.

Using Julia with the VS Code plugin

Julia code can also be saved and edited in files with a `.jl` extension. Numerous IDEs exist to do that. In this book, we'll use the VS Code platform with the excellent Julia plugin, which provides syntax highlighting and completion, lookup definitions, and plotting among many other features.

A handy way to start VS Code from the terminal prompt is by typing in `code`.

Search in the **Extensions** tab for Julia and install it. Then, open up a new file, and type in `println("Hi Web World from Julia!")`, and save it as `hiweb.jl`.

Run the program in VS Code with *F5* to see the string printed out. Or start the REPL and type the following to get the same result:

```
julia> include("hiweb.jl")
```

`include` evaluates the contents of the input source file.

Or, from a terminal prompt, execute a Julia source file simply by typing the following:

```
julia hiweb.jl
```

In all cases, the output will be Hi Web World from Julia!

The include command loads in the Julia file and executes the code.

Continuing the example from the previous section, if you want to start editing the newly created testset.jl file in VS Code when you are working in the REPL, simply type the following:

```
julia> edit(joinpath("testing", "phase1", "testset.jl"))
```

Now that you have some idea of working with Julia using the REPL, in package mode, and using VS Code, let us dig deeper into understanding the basics of the language. In the next section, we will explore some of the basic types, flow controls, functions, and methods in Julia.

Types, flow controls, and functions in Julia

In this section, we will discuss some basic concepts in Julia and start applying them to our ToDo project. Let us start by understanding the types of data that can be used in Julia.

Types

To achieve its high level of performance, Julia needs to know the types of data it will handle at either compile time or runtime. You can annotate a local function variable x with a type Int16 explicitly, like in x::Int16 = 42.

But you can just as well write x = 42. If you then ask for the variable's type with typeof(x), you get Int64 (or Int32 on 32-bit operating systems). So, you see, there is a difference: if you know Int16 is sufficient, you can save memory here, which can be important if there are many such cases.

Explicit typing is sometimes done for function arguments and can enhance performance. Types can also be added at a later stage of the project. Also, although Julia allows it, do not change a variable's type: this is very bad for performance. To test whether a variable is of a certain type, use the isa function: isa(x, Int64) returns true.

Julia has an abundance of built-in types, ranging from Char, Bool, Int8 to Int128 (and its unsigned counterparts, UInt8 and so on), Float16 to Float64, String, Array, Dict, and Set.

Strings containing variables or expressions can be constructed by *string interpolation*: when x has the value 108, the string "The value of x is $x" is evaluated to "The value of x is 108". An expression must be placed within parentheses, like "6 * 2 is $(6 * 2)", which evaluates to "6 * 2 is 12".

It is best practice not to use global variables as they cause bugs and have major performance issues. It is better to use constants, such as `const var1 = 3`, which can't be modified. In this case, Julia's JIT compiler can generate much more efficient code.

As an alternative to global variables, you can use `Refs` as is done in the Genie framework, like this:

```
const var = Ref{Float64}(0.0)
var[] = 20.0
```

That way, you make certain that the type of `var` will not change.

Types follow a hierarchy, with the `Any` type at the top, which, as the name says, allows any type for such a variable. In *Figure 1.2*, we show a part of this type tree:

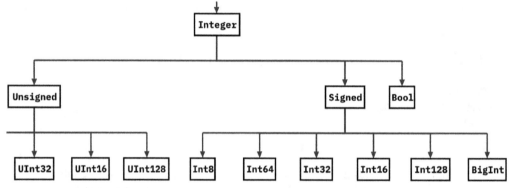

Figure 1.2 – Part of Julia's type hierarchy [Adapted from Type-hierarchy-for-julia-numbers.png made available at https://commons.wikimedia.org/wiki/File:Type-hierarchy-for-julia-numbers.png by Cormullion, licensed under the CC BY-SA 4.0 license (https://creativecommons.org/licenses/by-sa/4.0/deed.en)]

In the preceding figure, we see that the `Integer` type has subtypes `Unsigned`, `Signed`, and `Bool`.

A *subtype* (a kind of inheritance relationship) is indicated in code as follows:

```
Bool <: Integer
```

Types with subtypes are abstract types; we cannot create an instance of this type. The types that have no subtypes (the leaf nodes) are *concrete* types; only these can have data. For example, `Bool` variables can have the values `true` and `false`. A variable b declared as `Integer` has in fact the type `Int64`:

```
b :: Integer = 42
typeof(b)    # => Int64
```

To describe a ToDo-item, we need several data items or fields. Let us have a look at what types of values each field can take using some examples:

- **id**: Here, we could add an integer of type `Int32`, such as `1`.
- **description**: Here, we can only use a String, such as `"Getting groceries"`.
- **completed**: This field will take a `Bool` value, which is initially set to `false`.
- **created**: This field takes the `Date` type. This type lives in the `Dates` module, so to make it known to Julia, we have to say so in code: `using Dates`.
- **priority**: This field could take an integer between `1` to `10`.

We could group all this data into an array-like type, called a `Vector`. Because we have all kinds of items of different types, the type of the items is `Any`. So, our `Vector` would look as follows:

```
julia> todo1 = [1, "Getting groceries", false, Date("2022-04-
01", "yyyy-mm-dd"), 5]
```

Running the preceding code would give us the following output:

```
5-element Vector{Any}:
1
"Getting groceries"
false
2022-04-01
5
```

To get the *description*, we have to use an index, `todo1[2]`; the index is 2 because Julia array indices start from `1`.

A better way to group the data is using a struct:

```
julia> mutable struct ToDo
               id::Int32
               description::String
               completed::Bool
               created::Date
               priority::Int8
       end
```

Then, we can define the same `todo` item as in the preceding code as a `struct` instance:

```
julia> todo1 = ToDo(1, "Getting groceries", false, Date("2022-
04-01", "yyyy-mm-dd"), 5)
```

Now, instead of using an index, we can directly ask for a particular field, for example, the todo's `description`:

```
julia> todo1.description
"Getting groceries"
```

Or we can indicate when the item is dealt with:

```
julia> todo1.completed = true
```

To nicely print out the data of a struct, use the `show` (struct) or `display` (struct) functions.

Another thing that we will see used a lot in Genie is symbols. These are names or expressions prefixed by a colon, for example, `:customer`. Symbols are immutable and hashed by the language for fast comparison. A symbol is used to represent a variable in metaprogramming.

The `:` quote operator prevents Julia from evaluating the code of the expression. Instead, that code will be evaluated when the expression is passed to `eval` at runtime. The following code snippet shows this behavior:

```
ex = :(a + b * c + 1)
a = 1
b = 2
c = 3
println("ex is $ex")   # => ex is a + b * c + 1
println("ex is $( eval(ex) )")   # => ex is 8
```

See the *Useful techniques in Julia web development* section for how symbols can be used.

In this section, we have seen that the use of the appropriate types is very important in Julia: it can make your code more performant and readable.

Flow controls

Julia is equipped with all the standard flow controls, including the following:

- `if elseif else end`: Branching on a condition.
- `for in end`: Looping with a counter or iterating over a set of values.
- `while end`: Looping while testing on a condition.

- `break`: Used to jump out of a loop.

- `continue`: Used to continue with the loop's next iteration.

- `throw`: Used to throw exceptions and use code that can go wrong in a `try` construct. Here is an example:

```
try
# dangerous code
catch ex # handle possible exceptions
finally  # clean up resources
end
```

You can see a concrete usage example of `try`/`catch` in the echo server example in the *Making a TCP echo server with TCP-IP Sockets* section of *Chapter 2, Using Julia Standard Web Packages*. However, don't overuse this feature; it can degrade performance (for those curious, this is because the runtime needs to add the stack trace to the exception, and afterward, needs to unwind it).

- You can also make your own custom exceptions like this:

```
mutable struct CustomException <: Exception
# fields
end
```

Let's see an example of flow control in action. Here is how we compare the priorities of todos:

```
if todo2.priority > todo1.priority
    println("Better do todo2 first")
else
    println("Better do todo1 first")
end
```

So, you see, Julia has all the basic flow controls like any standard programming language.

Functions and methods

Functions are the basic tools in Julia. They are defined as follows:

```
function name(params)
 # body code
end
```

Alternatively, we can use a one-liner:

```
name(params) = # body code
```

Functions are very powerful in Julia. They support optional arguments (which provide default values when no value is provided) and keyword arguments (here the argument's `arg1` value must be specified as `func(arg1=value)` when the function is called). Functions can be nested inside other functions, passed as a parameter to a function, and returned as a value from a function. Neither argument types nor return types are required, but they can be specified using the `::` notation.

Values are not copied when they are passed to functions; instead, the arguments are new variable bindings for these values.

To better indicate that a function changes its argument, append `!` to its name, for example:

```
julia> increase_priority!(todo) = todo.priority += 1
julia> todo1.priority
5
julia> increase_priority!(todo1)
6
julia> todo1.priority
6
```

In the preceding code, notice that we don't need to indicate the type of the argument; `todo` functions are by default generic, meaning that in principle, they can take any type. The JIT compiler will generate a different compiled version of the function each time it is called with arguments of a new type. A concrete version of a function for a specific combination of argument types is called a method in Julia. You can define different methods of a function (also called function overloading) by using a different number of arguments or arguments with different types with the same function name.

For example, here are two overloading methods for a `move` function:

```
abstract type Vehicle end

function move(v::Vehicle, dist::Float64)
  println("Moving by $dist meters")
end
function move(v::Vehicle, dist::LightYears)
  println("Blazing across $dist light years")
end
```

The Julia runtime stores a list of all the methods in a **virtual method table** (**vtable**) on the function itself. Methods in Julia belong to a function, and not to a particular type as in object-oriented languages.

In practice, however, an error will be generated when the function cannot be applied for the supplied type. An example of such an error is as follows:

```
julia> increase_priority!("does this work?")
```

If you run the preceding code, you will get the following output:

```
ERROR: type String has no field priority
```

One could say that a function belongs to multiple types, or that a function is specialized or overloaded for different combinations of types. This key feature of Julia is called **multiple dispatch**, meaning that the execution can be dispatched on multiple argument types. Julia's ability to compile code that reads like a high-level dynamic language into machine code that performs like C almost entirely derives from this ability, which neither Python, C++, nor Fortran implement.

A function can also take a variable number of arguments, indicated by three dots (…, called the splat operator). For example, the validate function takes two arguments, a and b, and then a variable number of values (args…):

```
validate(a, b, args…)
```

The function can be called as validate(1, 2, 3, 4, 5), or as validate(1, 2, 3), or even validate(1, 2). The type of args… is Vararg; it can be type annotated as args::Vararg{Any}.

If you see what seems to be a function call prefixed with an @ (such as @error or @authenticated! in Genie), you are looking at a **macro** call. A macro is code that is modified and expanded at parse-time, so before the code is actually compiled. For example, @show is a macro that displays the expression to be evaluated and its result, and then returns the value of the result. You can see examples in action with @async in the *Making a TCP echo server with TCP-IP Sockets* section in *Chapter 2, Using Julia Standard Web Packages*.

In this section, we saw that Julia has pretty much what you expect in any modern programming language: a complete type system, normal flow controls, exception handling, and versatile functions. We cannot review all the methods that Julia has to offer in this book, but detailed information regarding methods can be found in the Julia documentation: https://docs.julialang.org/en/v1/.

Now that you know some basic features that define the Julia language, let us explore some useful techniques that can help us further to take advantage of the speed of Julia.

Useful techniques in Julia web development

In this section, we will highlight some methods and techniques that you will see used often in Julia web development and that we will also use in the project in *Part 2*. Here is a list of these key techniques:

- **Multi-line strings**: These strings are delineated with " " ". These are often useful in web apps to use chunks of HTML code in a variable as follows:

```
form = """ <form action="/" method="POST"
enctype="multipart/form-data"> <input type="text"
name="name" value="" placeholder="What's your name?" />
<input type="submit" value="Greet" /> </form> """
```

- **String substitution**: This can be used to insert variable contents into messages on the screen as follows:

```
<h4 class="container">
Sorry, no results were found for "$(params(:search_
movies))"
</h4>
```

- **do block syntax**: This syntax makes the code easier to read. Here is a simple example:

```
list = [1, 2, 3]
map(x -> x^2, list)
```

This can also be written as follows:

```
map(list) -> do x
    x^2
end
```

- Here is an example usage from the Genie framework:

```
route("/hello.html") do
  html("Hello World")
end
```

Here, route() and html() are functions from the Genie framework. The do x syntax creates an anonymous function with argument x and passes it as the first argument to the function stated before do. In this example, the string "Hello World" will be shown on the web page when the /hello.html URL is requested.

Another way to write this, which is clearly not that readable, is as follows:

```
route(html("Hello World"), "/hello.html")
```

We will be using this do syntax quite often in *Part 2*.

- &&: The Boolean *and* && operator is often used to write concise conditional code, for example:

```
isempty(strip(params(:search_movies)))
&& redirect(:get_movies)
```

The preceding code first evaluates the isempty(strip(params(:search_movies))) part; if this is false, nothing happens anymore. Only if the first part is true will the second, redirect(:get_movies), get evaluated. Thus, if the search_movies parameter has a value, only then will the redirect to :get_movies take place.

- ||: The *or* || operator doesn't evaluate the part after || when the first part is true, for example:

```
isa(getfield(m, field), Int) || return
ValidationResult(invalid, :is_int, "should be an int")
```

In the preceding example, if m is of type Int, the ValidationResult from the right-hand side is not shown.

- **Pipe operator**: This operator, denoted by |>, is quite handy. An example is as follows:

```
h1("Welcome Admin") |> html
```

The output of the preceding function before the pipe operator (|>) is given as a first argument to the function after the pipe. This allows for easy function chaining.

- **Ternary form**: An if condition a else b end statement is often written in a ternary form: condition ? a : b.

An example is as follows:

```
flash_has_message() ? """<div class="alert alert-
$flashtype alert-dismissable">$(flash())</div>""" : ""
```

If the function flash_has_message() returns true, then the multiline string after ? (which contains the HTML code for a div) is the result of the expression; if false the empty string "" is returned.

- **Symbols**: These are often used in Julia (web) code denoted by :, for example:

```
julia> sym = :info
:info
julia> typeof(sym)
Symbol
```

Symbols are used to indicate access to a variable (such as info), but what info exactly contains is not evaluated at that moment in code. A symbol gets replaced with the value bound to that symbol when the expression containing the symbol is evaluated at runtime.

Here is an example from Genie passing the values of the `:user_id` and `:user_status` variables in the payload to a `createUser` function:

```
createUser(postpayload(:user_id), postpayload(:user_
status, "active"))
```

So, `postpayload(:user_id)` can be seen as binding values to variables (something that can change) inside user-provided expressions. Here, on evaluation, `:user_id` is replaced by the value `Symbol` is pointing to.

- `<% %>`: Web frameworks often embed code inside HTML with `<% %>`. This can be used in the Julia Genie web framework, as follows:

```
<h4><% movie.description %></h4>
```

In the preceding code, `<% %>` is used to insert a `description` field of a `movie` instance in that place.

`<% %>` can also contain whole blocks of code as well as function calls.

Embedded code can also contain a call to the `@yield` macro, like in the following snippet from `app.jl.html` in Genie:

```
<body>
<div class="container">
<%
@yield
%>
</div>
</body>
```

`@yield` is used to output the content of the view/controller into the layout.

Do familiarize yourself with the techniques described in this section. Doing so will make Julia web code instantly more understandable. All Julia web apps are projects that contain modules and use packages, so that's what we'll discuss in the next section.

Using Julia modules and packages

Code in Julia is not limited to functions and can be organized at higher levels through modules and packages. A **module** is one level higher than functions under which code in Julia can be organized. A **package** is another level higher, can contain one or more modules, and provides functionality that can be reused by other Julia projects. Often, a web app is a package, containing a number of modules. When the package contains a project file called `Project.toml`, the file is also a **project**.

Modules

Modules are used to group together the definitions of types, functions, constants, and so on that are related. By convention, a file named `M.jl` will always define a module named `M`.

Such a module will be declared as follows (shown here for the Genie framework):

```
module Genie
# Loads dependencies and bootstraps a Genie app.
# Exposes core Genie functionality.
end
```

And it will be stored in `Genie.jl`.

To illustrate, let's create a module `ToDoApp` inside a `ToDoApp.jl` file, with the `ToDo` struct definition and a display function (see `Chapter1\modules\ToDoApp.jl` in the code repository):

```
module ToDoApp
using Dates   # to make the Date type available
export print_todo, ToDo

mutable struct ToDo
    id::Int32
    description::String
    completed::Bool
    created::Date
    priority::Int8
end

function print_todo(todo)
    if !todo.completed
        println("I still have to do: $(todo.description)")
        print("A todo created at: ")
        helper(todo)
    end
end

function helper(todo)
    println(todo.created)
```

```
        end
    end
```

In the preceding code, we see that `using` is needed to bring in the definitions of the `Dates` module.

In the REPL, we evaluate the preceding Julia script with `include(" ToDoApp.jl")`. Then, we employ `using .ToDoApp`.

The period (`.`) is used here because we want to look for definitions inside the scope of the current module. Without this, we get an error:

```
julia> using ToDoApp
ERROR: ArgumentError: Package ToDoApp not found in current path
Import
```

Also, we must do `using Dates` so that the `Date` type is recognized, which is needed when making a ToDo instance:

Now, let us define our struct instance as follows:

```
julia> todo1 = ToDo(1, "Getting groceries", false, Date("2022-
04-01", "yyyy-mm-dd"), 5)
Main.ToDoApp.ToDo(1, "Getting groceries", false, Date("2022-04-
01"), 5)
```

We can now call the *exported* `print_todo` function:

```
julia> print_todo(todo1)
I still have to do: Getting groceries
A todo created at: 2022-04-01
```

However, the helper function is not available because it was not exported from the module:

```
julia> helper(todo1)
ERROR: UndefVarError: helper not defined
```

But we can call the helper function as follows:

```
ToDoApp.helper(todo1) # => 2022-04-01
```

When evaluating using, Julia looks in the filesystem for modules or packages in paths that are stored in the LOAD_PATH variable. By default, LOAD_PATH contains the following:

```
julia> LOAD_PATH
3-element Vector{String}:
"@"
"@v#.#"
"@stdlib"
```

The preceding code implies that first the current project is searched, then the default Julia environment, and then the standard library.

The variable @__DIR__ contains the current folder. So, another way to enable Julia to search for modules or packages in the current folder is to say push!(LOAD_PATH, @__DIR__).

Let us summarize when and how to use using:

- using MyPackage looks in the LOAD_PATH for a file called MyPackage.jl and loads the module contained in that file; all exported definitions are loaded into the current scope

- using .MyPackage: This part of the code instructs looking for definitions inside the scope of the *current module*, which is needed because we have previously done include ("MyPackage.jl")

- using ..MyPackage: This part of the code instructs looking for definitions inside the *parent scope* of the current module

Besides include and using, we can also bring in a module with import. Then, you have to prefix the name of a function or another object with its module name when it is used. For example, after import Inflector has imported the Inflector module, you have to use its to_plural function, as Inflector.to_plural(name).

The import keyword also has to be used when you want to extend functions with new methods. For example, if you want to pretty-print your own types with a new version of the show function, you first have to do import Base.show.

To bring in specific definitions, use : after using or import as follows:

```
import SearchLight: AbstractModel
```

As an example, here are the starting lines of the Genie module:

```
module Genie
import Inflector
```

```
include("Configuration.jl")
using .Configuration
const config = Configuration.Settings()
include("constants.jl")
import Sockets
import Logging
using Reexport
Using Revise


# rest of the code
end
```

Packages and projects

Packages are managed using the Git version control system and the package manager, `Pkg` (which is itself a package!). They are stored on GitHub, and each Julia package is named with a `.jl` suffix. A single GitHub repository may host one or more packages, but a good convention is one repository containing just one package.

A single package with the name `P` will always contain a `P.jl` file. By convention, this is placed in a subfolder, `src`. You can't have other top-level modules in a single package.

As an example, the Genie framework, called `Genie.jl`, can be found at `https://github.com/GenieFramework/Genie.jl`.

A project is a package that contains two `.toml` files, which declare the packages your project depends on. You can create a project from the REPL as follows:

```
(@v1.8) pkg> generate MyPackage
Generating project MyPackage:
MyPackage\Project.toml
MyPackage\src/MyPackage.jl
```

The output will show the file structure created by the `generate` command.

The default `Project.toml` file contains the following:

```
name = "MyPackage"
uuid = "607adcac-db05-4b5b-9d7e-b11c396083d4"
authors = ["YourName<email-address>"]
version = "0.1.0"
```

A project can also contain a [deps] section, containing the names and universally unique ids (UUID) of the packages your project depends on (we will see an example of this in the next section). When adding a package to your project with the add command, the entry in the [deps] section is automatically filled in. The [compat] section constraints compatibility for the dependencies listed under [deps].

Besides Project.toml, a project can also have a manifest in the form of a Manifest.toml file, as indeed all Genie projects have. This file is generated and maintained by Pkg and, in general, should never be modified manually. The Manifest.toml file records the state of the packages in the current project environment, including exact information about (direct and indirect) dependencies of the project.

Given these two files, you can exactly reproduce the package dependency environment of a project, so this guarantees reproducibility.

Parsing a CSV file

As a simple example of how to work with packages, let's read the data from a CSV file and display it. Suppose we have a todos.csv file that contains a header line with column names, and then line by line, the field data of our to-dos, as follows:

```
id, description, completed, created, priority
1, "Getting groceries", true, "2022-04-01", 5
2, "Visiting my therapist", false, "2022-04-02", 4
3, "Getting a haircut", true, "2022-03-28", 6
4, "Paying the energy bill", false, "2022-04-04", 8
5, "Blog on workspace management", true, "2022-03-29", 4
6, "Book a flight to Israel", false, "2022-04-04", 3
7, "Conquer the world", true, "2022-03-29", 1
```

Start up a REPL to work with the data. We already installed the CSV package previously in the *Using the package mode to jump-start a project* section. We'll also need the DataFrames package to show our data as a table with columns, so go into pkg mode by typing] and give the command: add DataFrames.

Going back to the normal REPL, type the following using command:

```
using CSV, DataFrames
```

Now, we can read in the CSV file into a DataFrame object, df, with the following command:

```
df = CSV.read("todos.csv", DataFrame)
```

You will get the following output in the REPL:

```julia
julia> df = CSV.read("todos.csv", DataFrame)
7×5 DataFrame
 Row │ id     description                    completed  created     priority
     │ Int64  String31                       Bool       Date        Int64
─────┼───────────────────────────────────────────────────────────────────────
   1 │     1  Getting groceries                   true  2022-04-01         5
   2 │     2  Visiting my therapist              false  2022-04-02         4
   3 │     3  Getting a haircut                   true  2022-03-28         6
   4 │     4  Paying the energy bill             false  2022-04-04         8
   5 │     5  Blog about workspace management     true  2022-03-29         4
   6 │     6  Book a flight to Israel            false  2022-04-04         3
   7 │     7  Conquer the world                   true  2022-03-29         1
```

Figure 1.3 – Viewing a CSV file in a DataFrame

If the file has no header line, specify the `header=false` keyword argument. Also, if the data delimiter is something different, such as `;`, you can specify this with `delim=';'`.

The CSV package has a lot more capabilities for reading and writing, which you can learn about here: `https://csv.juliadata.org/stable/index.html`.

Now that you've seen how to use modules, packages, and projects, let's examine Julia's internal workings a bit more.

How Julia works

After this whirlwind tour of Julia, we want to see why Julia is a good fit for the web world. To do that, we must have a good understanding of Julia's internal workings.

The Julia JIT compiler works at runtime, starting from the Julia source code. Note that code from packages is most often precompiled. Type, method, and module definitions are written in an efficient serialized form so that the JIT can start compiling much faster. The first time a function is called with a certain combination of types of arguments, the correct machine code for those types is generated through **LLVM**. Moreover, the machine code is cached from then on, so after the initial compilation stage, the optimized code is looked up in the function's vtable (see *Types, flow controls, and functions*), and you get the bonus of much-improved performance.

Julia apps will often be long-running processes, so there needs to be a mechanism for freeing memory resources. The Julia developer, however, is not burdened with this task. Julia has a **garbage collector** process. This is a simple mark-and-sweep GC causing low overhead. The best advice here is as follows:

- Avoid unnecessary memory allocations
- Use standard-library methods that modify variables (whose name ends with `!`, for example, `sort!`) instead of creating new ones
- Use immutable objects (`const` and `struct`)
- Pre-allocate enough memory from the start to avoid GC altogether

Now that we know somewhat better how Julia works, we can argue why Julia can be used for web development.

Why Julia is a good fit for web development

Web development using Julia is very popular because of the following:

- The app needs to be deployed only on the server, not on an unknown number of clients, so the effort to put an application into production is minimal. This is true for all backend programming languages. To execute the app, you have the choice between installing a Julia runtime on your server or building a standalone executable with `PackageCompiler.jl`.
- Powerful web frameworks exist. This is certainly the case for main-stream dynamic languages, for example, **Ruby on Rails**, **Django** (a Python framework), or **Phoenix** (an Elixir framework). But Julia also saw the development of powerful web frameworks in recent years (we will discuss this more in *Part 2*), especially the Genie framework.
- Data scientists want to show the results of their research online in a visual and interactive fashion. Here, Julia, being a scientific and technical computing language, really shines.

Let's look at some important app properties and evaluate how Julia performs at these so that it becomes clear why Julia is a nice fit for web backend development and for exposing web services:

- **Speed**: Response speed is always of the utmost importance, no matter the kind of app you're running. Commonly used programming languages to develop web applications, such as PHP, Ruby, and Python, are interpreted languages. Typically, they are compiled to bytecode, which is deployed to a production machine where it is run on a **virtual machine** (**VM**). This VM translates the bytecode to machine code. This causes apps written in these languages to fall behind in certain benchmarks, so performance can be an issue.

Julia on the other hand is known for its excellent execution speed, because of its JIT compiler and highly optimized machine code generated through LLVM. That's why Julia often stays within the 2x range from optimized C code while outperforming dynamic languages with orders of magnitude (see `https://julialang.org/benchmarks/` and `https://benchmarksgame-team.pages.debian.net/benchmarksgame/fastest/julia-python3.html`).

Because of the JIT compiling stage, the startup of an application takes some time, which is sometimes called the **JIT latency**. So, often, the first execution of Julia code will take longer because execution waits for the compilation process. While this would be a problem for games or real-time apps, it is not an issue when developing web backends, which is a typically long-running process. Also, you can add a startup phase in which all your important code gets precompiled before exposing the app to users.

- **Garbage collection**: The freeing of memory through the GC will be less noticeable on a server. Its effect is spread out over all client processes. See the previous *How Julia works* section, for some advice here.

- **Scalability**: If your web app has an ever-growing number of users, you'll likely run into problems sooner or later. Julia is designed for concurrent/parallel and distributed execution, which makes it highly scalable and thus particularly suited for running massive apps with many users. If your web application requires heavy calculations or can come under a heavy load, Julia will make a great difference.

- **Platforms**: Because LLVM is used under the hood, Julia runs on very diverse platforms (see `https://julialang.org/downloads/#currently_supported_platforms`).

- **Functional**: Julia's emphasis on functions makes Julia a good choice for developing web services, which are typically function-oriented.

An ecosystem of libraries: Another of Julia's advantages over competing languages is that libraries can be combined and extended very easily. This allows for more code reuse, which means less time and effort is needed in Julia to develop a web app as in the competing language frameworks. Moreover, all Python libraries can be used via `PyCall.jl`, so in case no existing Julia package meets your need, you can use an appropriate Python library. The same goes for R with `Rcall.jl`, and Java/Scala with `JavaCall.jl`. Also, Julia can call C code directly, without any libraries needed.

We can conclude that Julia's performance and scalability characteristics and its extensive number of packages for visualizing data make it an excellent fit for the development of web apps, web services, and web dashboards.

Summary

In this chapter, we reviewed the Julia programming language in order to prepare ourselves for web development with Julia.

We worked with Julia in the REPL and with the VS Code editor, which is how we'll build web apps in the rest of the book. Then, we looked at types, flow controls, functions, and methods, which you'll need in any Julia app.

We followed that up with some useful Julia techniques in web development. We discussed modules and packages and illustrated them using the `CSV` and `DataFrames` packages.

Finally, we covered how the Julia runtime works and why Julia is a good fit for web development.

By now, you should be able to understand the underlying mechanisms of the code in future chapters and how to use Julia in your own projects.

In the next chapter, we'll dive into what Julia's standard library and **JuliaWeb** have to offer for building web apps.

Further reading

If you need a more thorough introduction or more details, follow a tutorial on `https://julialang.org/learning/tutorials/`, visit the Julia documentation at `https://docs.julialang.org/en/v1/`, or choose a book from `https://julialang.org/learning/books/`.

2
Using Julia Standard Web Packages

In this chapter, we will examine how the building blocks of the web, such as TCP-IP, WebSockets, the HTTP protocol, and more, are implemented in Julia's standard library, and we will show you how to build some functionality with these packages. Finally, we will apply what we have learned while working with JSON data.

Nearly all of the Julia web stack rests on the foundation laid by the **Sockets** package documented at `https://docs.julialang.org/en/v1/stdlib/Sockets/`, which we will use extensively in this chapter.

Building upon this, the Julia web stack lives at the JuliaWeb repo (`https://github.com/JuliaWeb/`), which contains a collection of packages for working with **Uniform Resource Identifiers (URIs)**, **Transport Layer Security (TLS)**, **Multipurpose Internet Mail Extensions (MIME)**, and more, all the way up to the HTTP package. Gathering nearly everything needed in the HTTP client and server functionality, we will use this package a lot in this chapter and the next.

In this chapter, we will cover the following topics:

- Making a TCP echo server with TCP-IP sockets
- Using WebSockets
- Building HTTP client and server functionality
- Working with JSON data using a REST web server

Technical requirements

To properly test the echo server, you'll need the *netcat* (*nc*) tool. By default, this is installed on Linux and macOS. For Windows users, you can get it from https://joncraton.org/blog/46/netcat-for-windows/.

The code in this chapter uses *HTTP v1.5*, *URIs v1.4*, and *WebSockets v1.5.10*.

We will also use Postman. There is a free version that you can download from `https://www.postman.com/downloads/`.

The source code for this chapter can be found at `https://github.com/PacktPublishing/Web-Development-with-Julia-and-Genie/tree/main/Chapter2`.

Making a TCP echo server with TCP-IP sockets

In this section, we will set up some foundational packages, then build a basic echo server, followed by a more sophisticated version.

The first version of an echo server

The **Sockets** package defines all the types (such as `IPAddr`, `IPv4`, `IPv6`, `TCPSocket`, and `UDPSocket`) and methods (such as `connect`, `listen`, `accept`, `bind`, `send`, and `recv`) for doing basic network communication in Julia.

Use the Sockets package to make IP addresses like this (see `Chapter2\tcp_echoserver\ip_addr_uri.jl`):

```
using Sockets
addr = ip"185.43.124.6"        # 1
typeof(addr)                   # 2
```

Line *1* uses the `ip"..."` string literal, and line *2* displays `IPv4`. Let's connect to a website and get its IP address:

```
connect("julialang.org", 80)   # 3
getaddrinfo("julialang.org")   # 4
```

Line *3* responds with the following:

```
TCPSocket(Base.Libc.WindowsRawSocket(0x00000000000003ec) open,
0 bytes waiting)
```

In comparison, line *4* gives you the IP address of the remote machine:

```
ip"151.101.130.49"
```

To work with URIs, you'll need to install and access the URIs package:

```
(@v1.8) pkg> add URIs
julia> using URIs, HTTP
```

Let us suppose req contains an HTTP request to https://www.google.com/search?q=mammoth like this:

```
req = HTTP.Messages.Request("GET", "https://www.google.com/
search?q=mammoth")
```

Then, you can extract the relevant info from the URL as follows:

```
uri = URI(req.target)                                        #
URI("https://www.google.com/search?q=mammoth")
uri.scheme # "https"
uri.host   # "www.google.com"
uri.path   # "/search"
uri.query  # "q=mammoth"
queryparams(uri)
# Dict{String, String} with 1 entry: "q" => "mammoth"
```

The UDP protocol has no built-in reliability. It is often used for fast, real-time applications where the loss of packets is not very critical. UDP works connectionless using the bind(), send(), and recv() methods (for more info, see https://www.cloudflare.com/learning/ddos/glossary/user-datagram-protocol-udp/).

On the other hand, the TCP protocol works reliably by establishing a duplex channel between the server and the client. It is reliable but slower than UDP and is used for critical applications such as payment gateways, downloading critical data files, and more. It uses the connect(), listen(), and accept() methods. For more info, see https://www.fortinet.com/resources/cyberglossary/tcp-ip.

Using the Sockets functionality, we can easily write the code for a TCP echo server (see Chapter2\tcp_echoserver\echoserver.jl):

```
using Sockets


server = listen(8080)              # 1
while true                         # 2
    conn = accept(server)          # 3
    @async begin                   # 4
```

```
    try                          # 5
      while true
        line = readline(conn)    # 6
        println(line)
        if chomp(line) == "S"    # 7
          println("Stopping TCP server...")
          close(conn)
          exit(0)
        else
          write(conn, line)          # 8 - the echo!
        end
      end
    catch ex                     # 9
      print("connection lost with error $ex")
      close(conn)
    end
  end # end coroutine block
end
close(conn)
```

The preceding server echoes back what the clients send to it, and it can also handle multiple clients. Start this server in its own terminal with `julia echoserver.jl`.

Here is how the server works:

- On line *1*, the `listen` method returns a `TCPServer` object listening on port 8080.

- On line *2*, an infinite loop is started to handle all the requests.

- The `accept` method on line *3* waits for an incoming client to make a connection. As soon as a client connects, a `TCPSocket` object called `conn` is made.

- The message from the client is read on line *6* and echoed back on line *8*. Again, this is done in a `while true` loop to continue receiving messages from that client. The server only stops listening to that client when the client disconnects.

- When any of the clients sends an `S` character, the server stops because of the logic starting on line *7*.

- This loop is enclosed within a `try catch` starting on line *5*. That's because, during network communication, several possible errors can occur.

- When an error occurs, it is bound to the ex object on line 9, where we can log the error message. For example, when a client terminal exits, you get a connection-ended error message: `Base.IOError("read: connection reset by peer (ECONNRESET)", -4077)`.

- Additionally, our code calls an @async macro on line 4. This @async macro starts a new coroutine to start the execution of the begin end block that comes right after it. The macro handles the connection with each particular client in a separate coroutine. This means that the @async block returns immediately after each new connection so that the server can continue accepting new connections through the outer while loop.

To simulate a client, we can use netcat (nc) or another equivalent tool. On the terminal, type in the following:

```
user@DESKTOP$ nc localhost 8080
Hello Server, can you hear me?          # client message
```

You'll see the following echo response from server:

```
Hello Server, can you hear me?
```

You can also simulate the client by using another REPL instance with code like this:

```
julia> using Sockets
julia> conn = Sockets.connect(8080)
TCPSocket(Base.Libc.WindowsRawSocket(0x00000000000002d0) open,
0 bytes waiting)
julia> write(conn, "Server, do you hear me?\n")
24
```

However, with the REPL option, the number of bytes that the server has received is returned, and the message is not properly echoed.

In the server console, you can read something like the following:

```
D:\Julia-Web-Development-with-Genie\Chapter2>julia echoserver.
jl
Server, do you hear me?
```

After the client disconnects, the following message appears:

```
connection ended with error Base.IOError("read: connection
reset by peer (ECONNRESET)", -4077)
```

Stop the server with *Ctrl + D*.

To start up the echo server script as a background job, type in the following command:

```
user@DESKTOP$ julia -q echoserver.jl &
[2] 274
```

In the preceding command, the -q option suppresses the Julia banner.

A more sophisticated version of an echo server

By using background tasks with @async, we can simulate the echo server in just one REPL session (see Chapter2\tcp_echoserver\echoserver2.jl):

```
# type these lines in the REPL:
using Sockets
errormonitor(@async begin                              # 1
  server = listen(8080)
  while true
    conn = accept(server)
    @async while isopen(conn)
      write(conn, readline(conn, keep=true))           # 2
    end
  end
end)


sleep(1) # give the server time to start
client = connect(8080)                                 # 3
errormonitor(@async while isopen(client)               # 4
  write(stdout, readline(client, keep=true))
end)
println(client, "Hello World from the Echo Server")    # 5
sleep(1) # give the server time to respond

close(client)                                          # 6
```

- On line *1*, we start a `Task` for the TCP server listening on port `8080`; `errormonitor` is just a `Base` function that prints an error log to a standard error output if the task fails. Essentially, this is a try-catch error handling mechanism (see the *Flow controls* subsection of *Chapter 1*, *Julia Programming Overview*) specifically for tasks, making the code much more readable. Julia responds with the following output:

```
Task (runnable) @0x000000000db9cf90
```

- We see the echo effect in the inner `while` loop on line *2*.
- Then, we connect a client to that port on line *3*, creating a `TCPSocket` object:

```
TCPSocket(Base.Libc.WindowsRawSocket(0x000000000000039c) open,
0 bytes waiting)
```

- Then, on line *4*, we start a `Task` for that client, which will echo anything sent to the client socket, to give the following output:

```
Task (runnable) @0x0000000008d41370
```

- On line *5*, we send a message to the client socket, and the server echoes that message:

```
julia> println(client,"Hello World from the Echo Server")
Hello World from the Echo Server
```

- Finally, on line *6*, the client is disconnected.

We have put to use the TCP protocol in Julia by writing a server that can handle multiple clients. Now, let's see how we can work with the WebSockets protocol in Julia.

Using WebSockets

WebSocket (`ws://URL`), or `wss://URL` for WebSocket Secure, is a bidirectional, low-latency, and high-speed protocol that builds on the TCP protocol. It uses HTTP(S) to make the connections (using ports `80` and `443`, respectively), so it also needs the `HTTP` package. In a way, it combines characteristics of both UDP and TCP; it is message-based like UDP but also reliable and stateful like TCP. Chat services are among the most popular applications of WebSockets.

You can read data from or write data to a WebSocket, send ping or pong messages, and close the connection.

The `WebSockets` package can be found at `https://github.com/JuliaWeb/WebSockets.jl`, which contains a complete example of the chat service.

To start using the `WebSockets` package, add it via the REPL's package mode:

```
(@v1.8) pkg> add WebSockets
```

The preceding command installs dependent packages such as URIs and HTTP and then precompiles them.

To set up a WebSocket, you need a URL and a callback function to run on the WebSocket.

Here is a simple program that sends a hundred pings to a crypto exchange site and measures the time it takes to receive the pong response (see `Chapter2\web_sockets\websockets1.jl`):

```julia
using WebSockets
function main()
WebSockets.open("wss://wsaws.okex.com:8443/ws/v5/
public")               do ws                                    # 1
            for i = 1:100
                    a = time_ns()                        # 2
                    write(ws, "ping")                    # 3
                    data, success = readguarded(ws)      # 4
                    !success && break                    # 5
                    b = time_ns()
                    println(String(data), " ", (b - a) /
                            1000000) #6
                    sleep(0.1)
            end
    end
end
main()
```

Let's break down how this works:

- The `open` command on line *1* takes the URL to connect to and constructs a `WebSockets` client in the do block, which specifies the callback function.

- `time_ns()` is a useful function from `Base` and is used to get the current time in nanoseconds.

- On line *3*, we write the `"ping"` string to the client.

- The `readguarded` function on line *4* gets the return data in the form of `Vector` and a `Bool` success variable.

- Line *5* breaks from the loop and exits the WebSocket handling when `success` is false, which means you can't write to it. When there is an answer, the `pong` data string is printed, along with the time it has taken in nanoseconds.

Here is an example output from the REPL:

```
julia> main()
pong 682.2282
pong 635.9462
pong 643.4656
pong 672.9338
pong 635.0898
pong 636.7789
```

You can also start the script from a terminal with `julia websockets1.jl`.

For the following example, we want to subscribe to a data feed to get the trade info (which is in JSON format) for XBT from the `https://www.bitmex.com` crypto trading site (see `Chapter2\web_sockets\websockets2A.jl`).

The JSON format

The JSON data format is the de-facto standard when it comes to data transport on the web, mainly because of its conciseness and readability. Additionally, it is used as a common design pattern when developing REST APIs, for example, to accept JSON payloads sent as `application/json` data over POST requests. We'll apply this in the *Working with JSON data using a REST web server* section. In *Part 2*, we'll see that the Genie framework has tailor-made functionality for this use case.

The Julia package ecosystem provides access to a couple of packages for working with JSON data. The most important ones are `JSON.jl` and `JSON3.jl`. `JSON.jl` was the default choice for many years, but now `JSON3.jl`, the more recent, more performant, and better maintained is becoming the default choice,

First, you have to install the JSON3 package with `(@v1.8) pkg> add JSON3`. Once the JSON3 package is added, run the following code:

```
# paste the following code in the REPL:
using WebSockets, JSON3
url = "wss://www.bitmex.com/realtime?subscribe=trade:XBT"

function open_websocket()
  WebSockets.open(url) do ws
    while isopen(ws)                        # 1
      data, success = readguarded(ws)
```

```
        success || break                    # 1B
        data = JSON3.read(String(data), Dict)   # 2
        print(data, "\n")
    end
    if !isopen(ws)                          # 3
        @async open_websocket()
    end
  end
end
@async open_websocket()
```

The preceding code should be easy to understand, as we have been through a similar example earlier in this section. However, the following list describes what's important in the preceding code:

- We read from the data feed in an infinite loop, testing on line *1* at each iteration whether the WebSocket is still open. The readguarded and writeguarded routines are socket read and write routines wrapped in try-catch. Besides the data, they return a success variable which is true when everything is ok. When an error has occurred, success is false and the break statement stops the execution.

- The data is returned is in the form of a JSON string, so we need to use the JSON3.read method on line *2*. We also pass the Dict type as the second argument of read to automatically convert the JSON data into a Julia dictionary.

- If the connection was lost, the socket is restarted on line *3*.

Here is a sample output:

```
julia> Dict{String, Any}("docs" => "https://www.bitmex.
com/app/wsAPI", "info" => "Welcome to the BitMEX Realtime
API.", "version" => "2022-04-01T10:20:40.000Z", "timestamp"
=> "2022-04-03T11:20:16.936Z", "limit" => Dict{String, Any}
("remaining" => 179))
Dict{String, Any}("success" => true, "request" => Dict{String,
Any}("args" => "trade:XBT", "op" => "subscribe"), "subscribe"
=> "trade:XBTUSD")
Dict{String, Any}("types" => Dict{String, Any}("grossValue" =>
"long", "homeNotional" => "float", …
```

We can make our code more flexible by storing the operation we want (in our example, we can use `subscribe`) and the trade we want to track (in our example, we can use `XBT`) in a `Dict` variable called `payload` (see line *1*). The code snippet for `Chapter2\web_sockets\websockets2B.jl` shows which code to use when you want to subscribe after the socket has been opened:

```
using WebSockets, JSON3
url = "wss://www.bitmex.com/realtime"
payload = Dict(                          # 1
             :op => "subscribe",
             :args => "trade:XBT"
          )

function open_websocket()
   WebSockets.open(url) do ws
     if isopen(ws)
       write(ws, JSON3.write(payload))    # 2
     end
# same code as above
    end
end
@async open_websocket()
```

In the preceding code, first, you'll have to `write` the `subscribe` JSON string (which contains the `payload` variable) to the WebSocket on line *2*.

You can run the code with `include("websockets2B.jl")`.

> **Note**
>
> Additionally, you might find the `SimpleWebsockets` package useful, which provides a more high-level but equally powerful interface to the WebSockets server and clients. Please refer to `https://github.com/citkane/SimpleWebsockets.jl`.
>
> The Genie framework (which is covered in *Part 2*) has its own built-in workflow for client-server communication over WebSockets. It is organized at a higher level than the code from the `WebSockets` package, using the *channel* concept: clients and the server exchange messages over channels.

The TCP and WebSockets protocols are basic ingredients of the web stack, but to really start building rich apps, you need an HTTP web server. That's exactly what we are going to implement in the next section, using Julia's HTTP package.

Building HTTP client and server functionality

The HTTP package is the workhorse of JuliaWeb and encompasses functionality from many previous packages. It contains a web server, an HTTP client, and also various utilities that make web development easier, such as the `Server`, `Router`, `HandlerFunction`, `Request`, and `Response` modules. You can find HTTP's full documentation at `https://juliaweb.github.io/HTTP.jl/stable/`. As always, get the package by typing `add  HTTP` into the REPL's package mode.

Using HTTP as a web client

An HTTP client sends an HTTP `Request` and returns an HTTP `Response`. Both are subtypes of `HTTP.Message`: `HTTP.Request` is an alias of `HTTP.Message.Request`. Here is an example of a fully populated `Request` object (see `Chapter2\http_client_server\request_response.jl`):

```
req = HTTP.Request(
"GET",          # 1 - Could be GET, POST, UPDATE and so on
"http://localhost:8081/search",    # 2 URL
["Content-Type" => "text/plain"],  # 3 Header fields
"Hi there!"                        # 4 Payload/body
)
```

Here is the code to access this URL (`http://localhost:8081/search`) data:

```
req.method          # "GET"
req.target          # "http://localhost:8081/search"
req.headers
# 1-element Pair{SubString{String}, SubString{String}}[]
Vector{Pair{SubString{String}, SubString{String}}}:
#   "Content-Type" => "text/plain"
# or more specific:
req["Content-Type"]   # "text/plain"
# same as: HTTP.header(req, "Content-Type")
HTTP.payload(req)
# 9-element Vector{UInt8}:
#   0x48
#   ...
#   0x21
String(HTTP.payload(req))
# "Hi there!"
```

The target is a complete URL, like `http://localhost:8081/search`. In such cases, the `splitpath` method of `HTTP.URIs` can be used to extract specific data, as shown in the following snippet:

```
splitpath(req.target)[2]          # returns "search"
```

The payload is a series of bytes, which must be explicitly converted with `String()`, such as `String(HTTP.payload(req))`.

`HTTP.Response` is an alias for `HTTP.Message.Response`.

Here is a Response object corresponding to the preceding request:

```
resp = HTTP.Response(
    200,               # 1 - Status code, 200 means success
    ["Content-Type" => "text/plain"], # 2 Header fields
    body="Hi there!"                  # 3 Payload/body
)
resp.status    # 200
resp.headers
# 1-element Vector{Pair{SubString{String}, SubString{String}}}:
#   "Content-Type" => "text/plain"
# or more specific:
resp["Content-Type"]    # "text/plain"
# same as: HTTP.header(resp, "Content-Type")
String(resp.body)            # "Hi there!"
String(HTTP.payload(resp)) # "Hi there!"
```

In the following code snippet, we show the `get` method in action (see how we abbreviated the output data for clarity in `Chapter2\http_client_server\http_client.jl`):

```
using HTTP
url = "https://julialang.org"
r = HTTP.get(url)
# HTTP.Messages.Response:
# """
# HTTP/1.1 200 OK
# ...
println(r.status) # 200
println(String(r.body))
```

```
# <!doctype html> <html lang=en > <meta charset=utf-8 >
# ...
```

In the preceding code, HTTP.get(url) is, in fact, a shortcut for HTTP.request("GET", url). HTTP.Response r is a struct containing the status, headers, and body. In the same way, the other HTTP verbs can be used, with HTTP.put, HTTP.post, and HTTP.head being defined as shorthand for HTTP.request("PUT", ...), and so on.

Setting up a web server with HTTP

In this section, we will show you how to set up a simple web server and add route handling.

A simple web server

The simplest way to set up a web server with HTTP is as follows (see Chapter2\http_client_server\http_webserver_simple1.jl):

```
# start in REPL                                        # or
with: julia htttp_webserver_simple.jl:
using HTTP
HTTP.listen() do http                          # 1
  while !eof(http)                             # 2
   println("body data: ", String(readavailable(http)))#3
  end
  HTTP.setstatus(http, 200)
  HTTP.setheader(http, "Content-Type" => "text/html")# 4
  HTTP.startwrite(http)                        # 5
  write(http, "ToDo 1: Getting groceries<br>")  # 6
  write(http, "ToDo 2: Visiting my therapist<br>")
  write(http, "ToDo 3: Getting a haircut")
end
```

Let us understand the preceding code:

- On line 1, a web server is started up to listen for HTTP connections; it will execute the do code block for every request. The http block variable is of the HTTP.Stream type.

- As long as data arrives at the stream (line 2), we can read from it (line 3).

- Then, we indicate on line 4 that we are going to send HTML content to the client, and we start writing some ToDo info on line 6.

Open up your favorite browser with the `localhost:8081` URL to see what the server has sent:

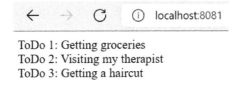

ToDo 1: Getting groceries
ToDo 2: Visiting my therapist
ToDo 3: Getting a haircut

Figure 2.1 – Output from the HTTP web server

By default, the web server will be located on **localhost** and port **8081**. In a more general scenario, you would write the following:

```
host = # ip-address in string format, like "127.0.0.1"
port = 8081
HTTP.listen(host, port) do http
          # code
end
```

If you want to see more info from the server, you can do `@show http.message`, which shows you the HTTP verb (here, it is GET), the host and port, the connection type, and much more. The HTTP headers can be viewed with `@show HTTP.header(http, "Content-Type")`.

A more request/response type of handling can be achieved with the `serve()` method (see `Chapter2\http_client_server\http_webserver_simple2.jl`):

```
using HTTP
HTTP.serve() do request            # 1
  try
                         # 2
    return HTTP.Response("Still lots of ToDos!")
  catch e
    return HTTP.Response(404, "Error: $e")
  end
end
```

In the preceding code, you can also see how to do exception handling with `try/catch`, starting on line *2*.

Setting up route handling

To handle routing requests, first, you make an HTTP.Router() object, such as the one on line *1* (see Chapter2\http_client_server\route_handling.j1), as follows:

```
using HTTP, Sockets
todos = """
ToDo 1: Getting groceries
ToDo 2: Visiting my therapist
ToDo 3: Getting a haircut
"""
const HOST = ip"127.0.0.1"
const PORT = 9999
const ROUTER = HTTP.Router()                              # 1
HTTP.register!(ROUTER, "GET", "/*", req -> HTTP.Response(200,
"Hello"))                         # 2
HTTP.register!(ROUTER, "GET", "/list_todos", req -> HTTP.
Response(200, todos))
HTTP.serve(ROUTER, HOST, PORT)                            # 3
```

Then, all of the routes have to be registered, one by one, with the register! function, starting on line *2*. Finally, instead of listen(), we use the serve() method, as shown on line *3*.

From the routing, we can see that /list_todos shows us a list of our to-dos, as shown in the following screenshot:

Figure 2.2 – Route handling

All other routes (such as /*) display Hello.

To test this out on your local machine, use Sockets.localhost instead of HOST. Also, notice that the root address, /, is not handled, so you get a page not found error 404 message. Again, you can use try/catch to show a more user-friendly message.

In this section, we saw how versatile the HTTP package is and how it forms the natural basis for a web server. Now we will build upon this and give the web server app some real functionality.

Working with JSON data using a REST web server

The HTTP package can also be used to expose Julia functions as a REST API, as we will see in this section. However, in order to tackle that, first, we need to know how to handle JSON data in Julia.

In the following examples, we'll use the JSON3 package.

Using the JSON3 package

In particular, the JSON3 package is widely used by the Genie framework. It is focused on speed by providing the following methods:

- JSON3.read (the equivalent of JSON.parse) to read a JSON string into a JSON3.Array (or a JSON3.Object). This allows you to access data with indexing or by using the dot notation, along with very flexible transformation to struct formats.

- JSON3.write which makes a JSON string.

Here is a code example that illustrates these two methods (see Chapter2\json_rest\json3_package.jl):

```julia
using HTTP, JSON3
todos = Dict( # type is Dict{Int64, String}
            1 => "Getting groceries",
            2 => "Visiting my therapist",
            3 => "Getting a haircut"
)
json_string = JSON3.write(todos)
# "{\"2\":\"Visiting my therapist\",\"3\":\"Getting a
haircut\",\"1\":\"Getting groceries\"}"
todos2 = JSON3.read(json_string)
# JSON3.Object{Base.CodeUnits{UInt8, String}, Vector{UInt64}}
with 3 entries:
#    Symbol("2") => "Visiting my therapist"
#    Symbol("3") => "Getting a haircut"
#    Symbol("1") => "Getting groceries"

todos2[1]   # "Getting groceries"
resp = HTTP.Response(
    200,
```

```
        ["Content-Type" => "application/json"],
            body=JSON3.write(todos)        # 1
)


# HTTP.Messages.Response:
# """
# HTTP/1.1 200 OK
# Content-Type: application/json
# {"2":"Visiting my therapist","3":"Getting a
haircut","1":"Getting groceries"}"""
body = HTTP.payload(resp) # 2 - a Vector of UInt8 bytes
io = IOBuffer(body)        # 3
todos = JSON3.read(io)     # 4
# or combined in one line:
todos = JSON3.read(IOBuffer(HTTP.payload(resp)))
# JSON3.Object{Base.CodeUnits{UInt8, String}, Vector{UInt64}}
with 3 entries:
#    Symbol("2") => "Visiting my therapist"
#    Symbol("3") => "Getting a haircut"
#    Symbol("1") => "Getting groceries"
```

In the preceding example, you can observe the following:

- On line *1*, we create a response using HTTP.Response with the todos data as the payload.

- We use application/json as the content type.

- The payload is always in string format, so we have to use JSON.write on line *1* to construct the body of the response. How do we get the data from this response? Using HTTP.payload in line *2* gives us a raw Vector of bytes.

- Then, applying IOBuffer on line *3* does the trick, and JSON.read on line *4* now returns a JSON3.Object (which is like a Dict variable) from the body data.

An **IOBuffer** is used in cases where you don't just want to output to a file, for example, if you want to preprocess the data before it is written.

Writing REST service endpoints for a ToDo app

Now, let us create a REST service web server that handles standard **create, read, update, and delete (CRUD)** operations through four different routes. We abbreviated the code somewhat (as indicated by . . .) to concentrate on what is important. The complete code can be found in the repo at https://github.com/PacktPublishing/Web-Development-with-Julia-and-Genie/blob/main/Chapter2/json_rest/webserver_json3.jl (see Chapter2\json_rest\webserver_json3.jl):

```julia
using HTTP, Sockets, JSON3, Dates
mutable struct ToDo    … end                          # 1
const ToDos = Dict{Int, ToDo}()                       # 2


function initToDos()       … end                      # 3


function getToDo(req::HTTP.Request)                   # 4
    todoId = HTTP.URIs.splitpath(req.target)[3]       # 4A
    todoId = parse(Int64, todoId)
    if haskey(ToDos, todoId)                          # 4B
        todo = ToDos[todoId]                          # 4C
        return HTTP.Response(200, JSON3.write(todo))# 4D
    else
        return HTTP.Response(200, JSON3.write("No ToDo with
          that key exists."))
    end
end


function deleteToDo(req::HTTP.Request)                # 5
    todoId = HTTP.URIs.splitpath(req.target)[3]
    todo = ToDos[parse(Int64, todoId)]
    delete!(ToDos, todo.id)                           # 5A
    return HTTP.Response(200)                         # 5B
end


function createToDo(req::HTTP.Request)                # 6
    todo = JSON3.read(IOBuffer(HTTP.payload(req)), ToDo)
                                                      #6A
```

```
        todo.id = maximum(collect(keys(ToDos))) + 1        # 6B
        ToDos[todo.id] = todo                              # 6C
        return HTTP.Response(200, JSON3.write(todo))       # 6D
    end


function updateToDo(req::HTTP.Request)                         7
    todo = JSON3.read(IOBuffer(HTTP.payload(req)), ToDo)
    ToDos[todo.id] = todo                                  # 7A
    return HTTP.Response(200, JSON3.write(todo))           # 7B
    end
```

In the preceding code, we can observe the following:

- The data model for this app is declared on line *1*; this is our ToDo struct definition from the *Types, flow controls, and functions in Julia* section in *Chapter 1, Julia Programming Overview*.

- On line *2*, we define a ToDos dictionary as a data store on the backend server, which we will populate with ToDo instances in the initToDos function (line *3*). This can evolve into a database backend during a more advanced stage of the app development.

- Lines *4* to *6* define the createToDo, getToDo, and other functions that run at the REST service endpoints. They all have Request as a parameter, taking out either of the following:

  - Part of the target path, such as getToDo or deleteToDo, to get the todoId of the requested ToDo

  - Its payload, such as createToDo and updateToDo

- In the getToDo function, we get todoId on line *4A* to fetch the struct in the dictionary: todoId = HTTP.URIs.splitpath(req.target)[3].

  The code assumes the URLs used in the app are of form /api/todos/7.

- We want to get todo7 out, as an example, so to get ID 7, we need index 3. We could have made this code less hardcoded with the last function, but we can try it as an exercise. See the following screenshot showing which index to use:

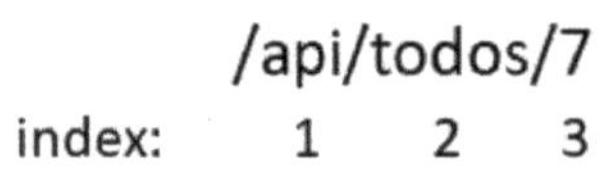

/api/todos/7

index: 1   2   3

Figure 2.3 – Index to use in the code

- Then, on line *4B*, we check whether we have a ToDo instance with the same ID as the haskey method. If that is the case, we fetch it on line *4C* and return it to the client on line *4D*. If not, we can send an error message: "No ToDo with that key exists.".

- The deleteToDo function (starting on line *5*) works almost entirely in the same way. The only difference is that we remove the ToDo instance from the dictionary with the delete! method on line *5A* and return an OK empty response on line *5B*.

- Looking at createToDo (starting on line *6*) in detail, we see that on line *6A*, the struct data of todo is extracted from the payload through the methods we saw in the *Using the JSON3 package* section: todo = JSON3.read(IOBuffer(HTTP.payload(req)), ToDo).

- On line *6B* of createToDo, we calculate the maximum of the ID key in our dictionary and add 1 as the key for the new Dict entry, which is made on line *6C*. Then, on line *6D*, this new entry is returned to the client.

- The updateToDo function (starting on line *7*) works almost entirely in the same way. The only difference is that we overwrite the ToDo instance from the dictionary on line *7A* and return an OK empty response on line *7B*.

Now comes the code, which defines the routing and starts up the web server:

```
JSON3.StructType(::Type{<:ToDo}) = JSON3.Struct()
```

For an explanation of this code and a deep dive into JSON3 StructType, please see the official documentation at https://quinnj.github.io/JSON3.jl/stable/#Struct-API:

```
initToDos()


const HOST = ip"127.0.0.1"
const PORT = 8080
const ROUTER = HTTP.Router()
HTTP.register!(ROUTER, "POST", "/api/todos", createToDo)#8A
HTTP.register!(ROUTER, "GET", "/api/todos/*", getToDo)  #8B
HTTP.register!(ROUTER, "PUT", "/api/todos", updateToDo) #8C
HTTP.register!(ROUTER, "DELETE", "/api/todos/*",
deleteToDo)                                             #8D
HTTP.serve(ROUTER, HOST, PORT)
```

In the preceding code, we can see the following occur:

- From lines *8A* to *8D*, we define the REST endpoints to dispatch to service functions explained previously.

For example, on line *8B*, we have the following:

```
HTTP.register!(ROUTER, "GET", "/api/todos/*", getToDo)
```

This code says that when the web server receives a GET request with a URL of form `api/todos/*`, then the `getToDo` method is called. The ending, `*`, is kind of a wildcard match, as it captures the path segment ID of the `ToDo` instance. For example, if called with `api/todos/42`, 42 is captured.

Looking at the route line of *8A*, we can see that we need to do a POST request for the URL ending with `"/api/todos"`.

As we can see from the `HTTP.jl` docs (see `https://juliaweb.github.io/HTTP.jl/stable/reference/#HTTP.request`), the POST request should be of the form `HTTP.request("POST", URL, [], "post body data")`.

In our case, the URL will be `http://localhost:8080/api/todos`.

- The last argument must be a `String`; in our case, it will be a JSON string because we want to read it with `JSON3`.

Now, let's do some tests to check whether the service works.

Testing the ToDo services in the REPL

We'll test the GET, DELETE, and POST requests, and leave you to test the PUT request.

For each request type, we give the REPL commands and the output that appears.

The get request

```
julia> using HTTP
julia> r = HTTP.get("http://localhost:8080/api/todos/3")
HTTP.Messages.Response:
"""
HTTP/1.1 200 OK
Transfer-Encoding: chunked
{"id":3,"description":"Getting a haircut","completed":true,"created":"2022-03-28","priority":6}"""
```

The delete request

```
julia> HTTP.delete("http://localhost:8080/api/todos/3")
HTTP.Messages.Response:
"""
HTTP/1.1 200 OK
```

```
Transfer-Encoding: chunked
"""
```

Then, we test whether `todo3` still exists:

```
julia> HTTP.get("http://localhost:8080/api/todos/3")
HTTP.Messages.Response:
"""
HTTP/1.1 200 OK
Transfer-Encoding: chunked
"No ToDo with that key exists.""""
```

The post request

In the REPL, do a `using Dates, JSON3` and define the ToDo struct, just like we did earlier on line *1*.

Now, create a new ToDo instance as follows:

```
todo8 = ToDo(8, "Listening music", false, Date("2022-04-07",
"yyyy-mm-dd"), 7)
```

Then, we convert it into JSON with the following:

```
json_string = JSON3.write(todo8)
"{\"id\":8,\"description\":\"Listening music\",\"completed\":fa
lse,\"created\":\"2022-04-07\",\"priority\":7}"
```

Finally, we do the POST request with the following:

```
HTTP.post("http://localhost:8080/api/todos", [], json_string)
```

As a response from the web server, our newly created ToDo instance is returned:

```
HTTP.Messages.Response:
"""
HTTP/1.1 200 OK
Transfer-Encoding: chunked
{"id":8,"description":"Listening music","completed":false,"crea
ted":"2022-04-07","priority":7}"""
```

To double-check this, we let the web server print out the complete dictionary after its creation (we left out some items, but show the added ToDo item with id 8):

```
# Dict{Int64, ToDo}(
5 => ToDo(5, "Blog about workspace management", true,
Date("2022-03-29"), 4),
8 => ToDo(8, "Listening music", false, Date("2022-04-07"), 7),
...
```

Testing the ToDo services with Postman

An easier way to test REST endpoints is to use Postman. This is software that is used to interact with web-based APIs. For example, here is the GET request from *The get request* section executed in Postman, showing the **Response** details at the bottom of the screen:

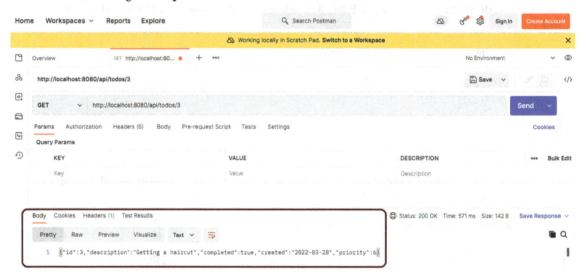

Figure 2.4 – The GET test using Postman

Testing the ToDo services with curl

You can also use the `curl` tool for testing from a terminal, as follows:

```
c:\>curl "http://localhost:8080/api/todos/3"
```

Then, the following output is displayed:

```
{"id":3,"description":"Getting a haircut","completed":true,"cre
ated":"2022-03-28","priority":6}
```

Note that `curl` is particularly handy for writing test scripts.

Summary

In this chapter, we've dived into what Julia's standard library and JuliaWeb have to offer for building web apps. First, we looked at working with the `Sockets` package, and we built two versions of a TCP echo server.

Then, we examined how to work with the `WebSockets` package. We followed that up with a thorough discussion of the `HTTP` package, starting by using it as a web client. Next, we built several web server examples and saw how to route incoming requests.

Finally, it all came together when we looked at how to work with `JSON` data using a `REST` web server. This was just a basic example, but I hope you're starting to see the beginning of a microservices architecture here. Armed with this knowledge, you'll be able to build your own service apps.

In the next chapter, we will discuss how Julia is used in various web use cases.

Further reading

If you need to refresh your knowledge of just about any web technology, go and visit the excellent reference site by Mozilla, which is located at `https://developer.mozilla.org/en-US/docs/Web`.

Also, on the Julia Discourse chat channels, there is a place called *Web Stack* (`https://discourse.julialang.org/c/domain/web/28`) that is specifically dedicated to answering questions about all web topics and Julia.

3
Applying Julia in Various Use Cases on the Web

In this chapter, we will discuss the higher-level packages and frameworks that exist in Julia to support web app development. Some of these include the *Julia website* (`https://julialang.org/`) built with *Franklin*, a static website generator; the Dash framework for creating dashboard apps (see the app gallery at `https://dash.gallery/Portal/`); and the *Genie MVC* web framework at https://www.genieframework.com/index.html#app-gallery. Additionally, the *JuliaGizmos* collection (`https://github.com/JuliaGizmos`) represents strong examples of cutting-edge development.

This chapter will give you an overview of what has been achieved in Julia using various web technologies, such as building static websites, microservices apps, dashboards, and web frameworks. By the end of the chapter, you'll have a better idea of which frameworks to use and gain a head start in applying these packages to your own projects.

In this chapter, we will cover the following topics:

- Using Franklin to build static websites in Julia
- Using Julia with **JavaScript (JS)**
- Building microservices with Julia
- Building dashboards with Julia
- Compiling Julia in WASM
- Using cloud services with Julia
- Choosing the right Julia framework for a project

Technical requirements

The source code for this chapter can be found at `https://github.com/PacktPublishing/Web-Development-with-Julia-and-Genie/tree/main/Chapter3`.

Using Franklin to build static websites in Julia

The *Franklin* package (see `https://franklinjl.org/`) provides a very nice framework for generating static websites. It is especially suitable for all kinds of technical blogging. It can render *LaTeX* and *KaTeX* math expressions and embed Julia code with live evaluation. The package is easy to use, light, and well documented. To get started with using Franklin, perform the following steps:

1. Go into REPL pkg mode with] and type in `add Franklin`.

2. Then, in normal REPL mode, type in `using Franklin`.

3. When generating a new website, first, you have to choose the template on which your website will be based. Navigate to `https://tlienart.github.io/FranklinTemplates.jl/` and choose one. All of the themes are written in *CommonMark* and can be adapted according to your needs. The generation phase converts the Markdown into HTML.

    Next, open the REPL with `julia -project` and issue the `newsite` command along with a name for your website and a chosen template (for example, the `vela` theme):

    ```
    julia> newsite("JuliaWebDev", template="vela")
    ```

    The preceding code will produce the following output:

    ```
    ✓ Website folder generated at "JuliaWebDev" (now the
    current directory).
    → Use serve() from Franklin to see the website in your
    browser.
    ```

    The newly generated website folder contains the HTML, JS, and CSS assets you need for this template. Additionally, the folder includes a `Project.toml` file containing all your project's dependencies, along with a number of `.md` files (such as `index.md`, `menu1.md`, and more), which you can modify to better suit your needs (I've modified some contents of `index.md` to better suit this book's subject).

4. Now, let's see our newly created website by issuing the `serve()` command:

    ```
    julia> serve()
    ```

    I. You will see the following output displayed:

    ```
    Activating project at `D:\Julia_Docs\Book\Julia-Web-
    Development-with-Genie\Chapter3\JuliaWebDev`
    → Initial full pass...
    ```

```
→ Starting the server...
√ LiveServer listening on http://localhost:8000/ ...
  (use CTRL+C to shut down)
```

Generating the preceding output takes around 10 seconds, but this much time is also needed to fully generate the site at __site. Don't change anything in the __site folder because it will be overwritten by your next modification!

Your folder structure now looks like the following:

.github	4/14/2022 8:53 AM	File folder	
__site	4/14/2022 9:17 AM	File folder	
_assets	4/14/2022 8:53 AM	File folder	
_css	4/14/2022 8:53 AM	File folder	
_layout	4/14/2022 8:53 AM	File folder	
_libs	4/14/2022 8:53 AM	File folder	
_rss	4/14/2022 8:53 AM	File folder	
.gitignore	4/14/2022 8:53 AM	Git Ignore Source File	1 KB
.gitlab-ci.yml	4/14/2022 8:53 AM	Yaml Source File	1 KB
404.md	4/14/2022 8:53 AM	Markdown Source File	1 KB
config.md	4/14/2022 8:53 AM	Markdown Source File	1 KB
index.md	4/14/2022 9:31 AM	Markdown Source File	9 KB
menu1.md	4/14/2022 8:53 AM	Markdown Source File	4 KB
menu2.md	4/14/2022 8:53 AM	Markdown Source File	5 KB
menu3.md	4/14/2022 8:53 AM	Markdown Source File	4 KB
Project.toml	4/14/2022 8:53 AM	Toml Source File	1 KB

Figure 3.1 – The Franklin folder structure

II. The serve() command also automatically starts a browser on port **8000** to view the landing page (which is derived from index.md) of your new website:

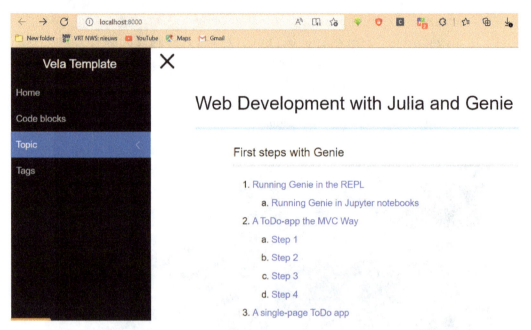

Figure 3.2 – The starting page of a Franklin website

The preceding screenshot shows a structured base page, with a collapsible menu on the left-hand side.

5. Open your editor in `index.md` or the other asset pages alongside the website in your browser. Any change you make will be immediately visible – how neat is that!

    You can also make any subfolder pages such as `pages/page1.md` to further complete your website.

6. Stop the web server with *CTRL + C*.

If you want to continue working on your website after having closed the REPL earlier, open up a terminal session in the website's root folder (which is `JuliaWebDev` in our case). Then, open the REPL and type in the following:

```
using Franklin
serve()
```

> **Franklin comes with some handy tools:**
>
> `verify_links()`: This checks whether all your hyperlinks are still working.
>
> `optimize()` : This minifies all HTML/CSS and preloads the built-in JS libraries.
>
> `publish()`: This deploys your website to a remote repo that is synced with your local website.

To test them, open the REPL in the folder with `julia -project`, and then issue a `using Franklin` command.

However, in many cases, you need more than just a static website. Don't worry, though, as Julia has you covered. Let us find out how by exploring how Julia can interact with *JS*.

Using Julia with JavaScript

In certain use cases, you might want to use *JS* for client-side web programming together with Julia. The client JS code could even be generated by Julia using part or a combination of the `JSExpr`, `Blink`, and `WebIO` packages, as we'll see in the following sections.

JSExpr

The `JSExpr.jl` package (`https://github.com/JuliaGizmos/JSExpr.jl`) translates Julia syntactically into JS. As usual, you can get it in the REPL package mode through `add JSExpr`, and then a `using JSExpr` command in the normal REPL.

JSExpr works by providing two dedicated macros: `@js` and `js""`.

`@js` transpiles a Julia expression into JS. It does this by making a `JSString`, which contains the JS code as a string. The following is an example of converting a Julia cube function into its JS equivalent:

```
@js function cube(arg)
return arg * arg * arg
end
```

The return value is as follows:

```
JSString("function cube(arg){return (arg*arg*arg)}")
```

In the following example, we see how `@js` also interpolates values and then translates an anonymous function into a JS function:

```
var1 = 108
callback = @js n -> n + $var1
```

The preceding code returns the following output:

```
JSString("(function (n){return (n+108)})")
```

The `js""` macro can also perform interpolation of Julia values into the string, for example, take a look at the following:

```
message = "hi"
fun1 = js"
function () {
            alert($message) // you can interpolate Julia
                            // variables!
}
"
```

The preceding code returns the following `JSString`:

```
JSString("function () {\n    alert(\"hi\") // you can
interpolate Julia variables!\n}\n")
```

The preceding `JSString` can be evaluated in JS.

As you can see, both macros generate `JSStrings` that can then be sent over a network socket to be interpreted in a browser. The Julia code to be converted already contains JS functions such as `console.log()`, `alert()`, or `document.querySelector()`.

`JSExpr` can be used to run the same data validation logic on the client and server sides. `JSExpr`, in itself, is fairly limited because it only offers execution of very simple Julia functions, as defined with the `@js` macro, or the execution of JS strings.

`JSExpr` does not provide interactivity between Julia and JS. If you need that, check out the `WebIO` package in the next section. `JSEXpr`, combined with `WebIO` to manage communication and execution, enables quite a lot of use cases.

Blink

Blink is Julia's wrapper for *Electron*, which is a framework to write desktop apps using web technologies (see `https://www.electronjs.org/`).

The source repo for Blink is located at `https://github.com/JuliaGizmos/Blink.jl`.

As usual, get it in the REPL package mode through `add Blink`, and then type `using Blink` in the normal REPL. If Electron is not installed on your machine, you can do so with `julia> Blink.AtomShell.install()`.

Blink displays HTML content on local web pages and makes Julia communicate with these pages through web sockets. Combining the strengths of Electron and Julia, Blink is effectively a toolkit for building web applications on the desktop.

Let's play a little bit with Blink.

In the REPL, do a using Blink command, and then type in w = Window(). This starts up Electron and opens up a new desktop window with the Julia icon and name as its title. Now you can add content to the window as follows:

```
body!(w, "Hello from Blink!")
```

The preceding code will generate the following window:

 Julia

Hello from Blink!

Figure 3.3 – The generated Blink window

To show external web pages, use the loadurl(w, url) function, loadurl(w, "https:// google.com").

Use the load! function to execute standalone HTML, CSS, or JS files. The next section contains some more advanced examples because Blink and WebIO are very intertwined.

WebIO

WebIO (see the repo at https://github.com/JuliaGizmos/WebIO.jl) functions as a bridge between Julia and the web because it provides useful high-level tools for communication between Julia and the browser. It displays and can interact with web content such as the *Jupyter Notebook*, *Blink* (see the preceding section), and *Mux* (see the *Mux* subsection under the *Choosing the right Julia framework for a project* section).

If you installed JSExpr earlier in this section, you should already have WebIO. If not, get it in the REPL package mode through add WebIO, and then type using WebIO in the normal REPL.

Now you can start building up an HTML page with the dom function:

```
body!(w, dom"p"("Hello from WebIO and Blink!"))
```

(On macOS and Ubuntu, you might also need to have Interact installed; see the next section.)

Using the @js and js "" macros from JSExpr allows you to execute JS, as we can see in the following code:

```
body!(w, dom"button"(
    "Greet",
    events=Dict(
        "click" => js"function() { alert('Hello, World!'); }",
```

```
    ),
  ))
```

The preceding code displays an HTML button with the text of `Greet` and wires in a JS `click` event. The event generates an alert pop-up window, as shown in the following screenshot:

Figure 3.4 – Alert popup in the Blink window

WebIO is quite rich in terms of features; discussing all of its functionality in this chapter is impossible, so consult the docs at `https://juliagizmos.github.io/WebIO.jl/latest/` and the *Mux* section.

Integration with NodeJS

Maxime Mouchet has developed an example framework for the use case when you want to deploy a Julia algorithm(s) in a `Node.js` environment. Why would you do this? Here is why:

- Julia is focused on developing performant algorithms
- Node.js is commonly used for writing web services in JS

The solution is based on the following steps:

- Make a Julia system image, which includes the algorithm's compiled code, by using `PackageCompiler.jl`.
- Build a Node.js native extension in *C/C++* in order to make the Node.js runtime and Julia work together. This will serve as a bridge between the Node.js app, which is written in JS, and the Julia code.

For more details, see the repo at `https://github.com/maxmouchet/julia-node-extension-demo` and the video of the talk given on this subject at *JuliaCon 2020* (`https://www.youtube.com/watch?v=bk3DNy18c0Y`).

Another approach to integrating Node.js and Julia can be seen at `https://github.com/waTeim/node-julia`.

In *Chapter 2, Using Julia Standard Web Packages*, under the *Working with JSON data using a REST web server* section, we saw how to write a simple microservice architecture using only the HTTP package. In the next section, we will discuss some interesting microservices projects that build upon these concepts.

Building microservices with Julia

Microservices are a well-established web technology used to deliver specific functionality over the web. They enable, among other benefits, easier deployment, management, and scaling. Julia is a good fit for creating an efficient environment for microservice development, as we will demonstrate in the different projects discussed in this section. A typical use case would be an analytics application that processes data coming from different sources over the internet. Such a microservice could further perform analysis, store results in a database, and show these results to end users through an on-demand microservice application.

In the *Integration with Node.js* section, we saw an example of how to integrate Julia into existing JS environments. This integration makes it possible to write microservices with Julia.

Another complete application, written by Jacob Quinn, uses a typical **model-view-controller** (MVC) architecture for microservices, which is not Julia-specific. It was presented as a workshop during *JuliaCon 2020* (see `https://www.youtube.com/watch?v=uLhXgt_gKJc&t=207s`), with the source code stored at `https://github.com/quinnj/MusicAlbums.jl/`. As the `project.toml` file shows, the app only uses the HTTP and JSON3 packages to implement the network functionality. All Julia files reside in one `src` folder. `MusicAlbums.jl` is the main code file, and the app starts with the following `run()` function:

```
server = @async MusicAlbums.run(DBFILE, AUTHFILE)
```

The app contains an authentication module in `Auth.jl` and a `loginUser` function in `Client.jl`. The data is contained in the `SqLite` database, `albums.sqlite`. The model is defined in `Model.jl`, which contains the mutable `Album` and `User` structs that are used for authentication.

The `Resource.jl` module contains the `createAlbum`, `getAlbum`, `updateAlbum`, `deleteAlbum`, and `pickAlbumToListen` microservice functions. Their route handling is defined here with the `HTTP.@register` macro. The `Service.jl` module details the microservices functions, which uses the `Mapper` module to translate to SQL.

Download the project's code and follow along with the workshop video (`https://www.youtube.com/watch?v=uLhXgt_gKJc&t=207s`) to see how the app can be built in 9 steps.

Another talk given at *JuliaCon 2021* (`https://live.juliacon.org/talk/SHHKEM`) shows you how to develop an SDK for the easier development of microservices and how it can be applied in a corporate environment.

In *Part 2* of this book, we will see how Genie makes writing microservices with Julia really easy. A key area of interest in using Julia for web apps comprises interactive data dashboards. Next, we will explore how to build these dashboards using `Dash`.

Building dashboards with Julia

Data scientists want to build rich, analytic, elegant, and exploratory portals to their data. In *Part 2*, we'll explore how to build a dashboard with the Genie framework (see *Chapter 7, Creating an Interactive Data Dashboard*). However, in this chapter, we'll concentrate on the *Dash for Julia* project, which was first published in October 2020.

Dash consists of two components that communicate via a *JSON* interface:

- The client side (frontend), written in JS, that is executed in the browser and contains the bulk of the code, control, and complexity. This makes `Dash` apps cross-platform and mobile-ready. The frontend mainly uses `Plotly.js` and the *React* ecosystem. It enables the direct binding of modern UI elements such as drop-down lists, sliders, and graphs to analytical Julia code on one or more web pages.

- The server side (backend), written in Julia, leverages the `HTTP.jl` package (there are also backends for *Python*, *R*, and *F#*).

The main advantage of using Dash is that the data scientist only needs to write Julia code in their project.

To see an example of visualizing COVID-19 data in U.S. states, navigate to `https://covid-county-dash.herokuapp.com/` (the source code can be found at `https://github.com/mbauman/CovidCountyDash.jl`) or visit the app gallery at `https://dash.gallery/Portal/`.

As usual, install the `Dash` package via `add Dash` in the REPL package mode.

Let's build a simple example. We'll plot gender against some flavor preferences in a bar graph. The following is the code you need (see `Chapter3\using_dash.jl`):

```
using Dash
app = dash()                             # 1
app.layout = html_div() do               # 2
    html_h1("Hello from Dash!"),
    html_div("Dash: A web application framework for your
             data."),
    dcc_graph(
```

```
        id = "example-graph-1",
        figure = (
            data = [
                (x = ["Chocolate", "Strawberry",
                "Vanilla"], y = [21, 19, 33],
                 type = "bar", name = "Male"),
                (x = ["Chocolate", "Strawberry",
                 "Vanilla"], y = [38, 18, 12],
                 type = "bar", name = "Female"),
            ],
            layout = (title = "Flavor Preferences by
                     Gender", barmode="group")
        )
    )
end
run_server(app, "0.0.0.0", debug=true)    # 3
```

The preceding code works as follows:

- Line *1* starts up the Dash framework and returns an application object
- Its layout field is described, on line *2*, as a tree of components, such as html_div, html_h1, and dcc_graph
- Line *3* starts up a web server, by default, on port 8050

Dash is declarative: each component is described entirely by keyword attributes. You can run the code in the REPL, but running it in a terminal is far more enjoyable because it enables hot reloading.

Start your app in the terminal with the following:

```
D:\Chapter3>julia using_dash.jl
[ Info: Listening on: 0.0.0.0:8050
```

Then, simply make any change in the code and save it, and the browser refreshes automatically!

The following is the output when you start a browser with the `http://127.0.0.1:8050/` URL:

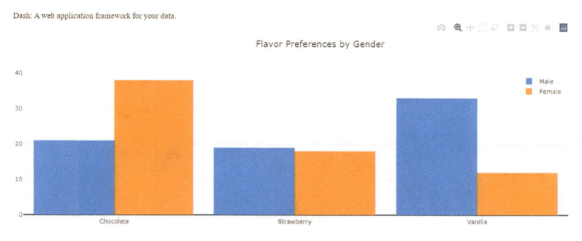

Hello from Dash!

Dash: A web application framework for your data.

Figure 3.5 – A Dash graph

Of course, `Dash` has many more possibilities; to find out more, consult the docs for this project at `https://dash.plotly.com/julia`.

Now that we have seen that Julia has a very viable dashboard builder with `Dash`, let's see what has happened with Julia in the emerging WebAssembly field.

Compiling Julia in WASM

WebAssembly (**WASM**) is a new, upcoming technology. It defines a binary code format to be executed in a stack virtual machine at near-native code execution speeds. WASM works in browsers, where it is hosted in the JS environment to replace computation-intensive JS code in applications such as games. However, WASM can also be deployed in standalone runtime environments. More than *40* programming languages use (or are trying to use) WASM as a compilation target, and Julia is no exception. Generating WASM from Julia opens up many opportunities to embed Julia apps in web interfaces. Julia has an advantage here: it generates machine code through *LLVM*, and LLVM supports WASM, so there seems to be a straightforward path from Julia to WASM.

A lot of work has been done in this area by *Tim Short* (see `https://github.com/tshort/ExportWebAssembly.jl`); his project provides a working prototype. Another experimental compiler, called *Charlotte*, was built by *Mike Innes* (see `https://github.com/MikeInnes/Charlotte.jl`). Here is an online playground where you can try out conversions of Julia into WASM and JS: `http://julia2js.gotfork.net/charlotte`.

As a trivial example, type in the following `add` function in the playground (notice that you have to indicate the types on the left-hand side):

Convert Julia to WebAssembly

Write Julia code to define a function in the box below and supply the name of the see if we can convert it WebAssembly. If you're not feeling creative you can use the

You can read more about what's going on at the GitHub project page.

Julia inputs

Function name & argument types:

add

Int, Int

```
1 function add(x, y)
2   x + y
3 end
```

Figure 3.6 – Input for the WebAssembly conversion

Then, push the **Convert to WebAssembly!** button and you get the generated WebAssembly in text (`.wat`) and binary (`.wasm`) format:

Julia to JS   ExportWebassembly.jl   Charlotte.jl

CONVERT TO WEBASSEMBLY!

Success!

Looks like you have some WebAssembly. It's shown below both in the text format as well as the binary format (following the example here).

WebAssembly text format

```
1 (module
2   (export "add" (func $#add_Int64_Int64))
3   (func $#add_Int64_Int64  (param i64) (param i64) (result
4     (get_local 0)
5     (get_local 1)
6     (i64.add)
7     (return)))
```

WebAssembly binary build log

```
1 0000000: 0061 736d
2 0000004: 0100 0000
3 ; section "Type" (1)
4 0000008: 01
5 0000009: 00
6 000000a: 01
7 ; type 0
```

Figure 3.7 – Output of the WebAssembly conversion

The two most recent exploratory projects building on *Short's* and *Innes'* work are the following:

- A proof-of-concept WASM port of the complete Julia runtime by *Keno Fischer* (`https://github.com/Keno/julia-wasm`). The WASM port is demonstrated by running the Julia REPL in the browser as WASM code. Additionally, it contains a setup of instructions for building Julia to WASM by yourself.

- The `WebAssembly.jl` package by *Mike Innes* (`https://github.com/MikeInnes/WebAssembly.jl`) contains tools for working with the WASM format in Julia.

To summarize, although a lot of groundwork has been done, at the time of writing this book, there isn't a production-ready tool in the Julia ecosystem to support WASM fully. Probably such a solution has to wait until the Julia compiler can support minimal binary executables (that is, one executable of minimal size containing the app and the Julia runtime), and also until WASM has better support for **garbage collection** (**GC**) because Julia needs GC.

Next, we will take a look at how we can use cloud services for and from Julia.

Using cloud services with Julia

As a beginning web developer, you'll start with the local development of a web or microservices application, and then progress to deploy the app to virtualized cloud environments. This section will tell you what cloud computing possibilities exist for Julia.

Deployment to cloud platforms

From the start, Julia was meant to be a language for distributed and cloud computing. *Julia Computing*, which is the company behind Julia, offers its own cloud computing solution named *JuliaHub* (`https://juliahub.com/lp/`) for academic and enterprise users. If you're looking for an alternative to the former free *JuliaBox* site, consult `https://github.com/xiaodaigh/awesome-data-science-notebook-engines`.

All of the major commercial cloud providers support deploying Julia apps. The *JuliaCloud* website (`https://github.com/JuliaCloud/`) contains most of the packages needed in this space.

The following is a list of the major ones:

- **Amazon Web Services** (**AWS**): AWS (`https://aws.amazon.com/`) is probably one of the biggest and best-known providers of cloud services and still offers the *AWS Free Tier* option where you can try it out. Follow the tutorial at `https://aws.amazon.com/getting-started/guides/deploy-webapp-apprunner/` to get started.

AWS.jl (https://github.com/JuliaCloud/AWS.jl) is Julia's interface package for AWS, supporting low-level and high-level APIs. The package has two backends, which serve as HTTP clients to reach the AWS REST API. You can consult the complete docs at https://juliacloud.github.io/AWS.jl/stable/. For a usage example, see using_AWS.jl.

- **Google Cloud Platform** (**GCP**): The GCP services (https://cloud.google.com/) are a good alternative to AWS. Their storage API can be interfaced with the GoogleCloud.jl package. You need to have a Google account to access the web-based management console at https://console.cloud.google.com/ and create a GCP project. See the tutorial in the docs at https://juliacloud.github.io/GoogleCloud.jl/latest/ for an introduction to the Google Cloud Service API.

- **Microsoft Azure**: The Microsoft cloud platform (https://azure.microsoft.com) is covered by the Azure.jl package (see the repo at https://github.com/JuliaComputing/Azure.jl). Moreover, Azure provides data science virtual machines with Julia and Jupyter preinstalled (https://azure.microsoft.com/en-us/services/virtual-machines/data-science-virtual-machines/). Julia microservices apps have successfully been deployed to *Azure Service Fabric*.

- **Digital Ocean platform**: For this platform, there is an interface package called JuDO.jl (https://github.com/aliddell/JuDO.jl).

Deployment to container platforms

Cloud services in Julia can be deployed through the **Docker** or **Kubernetes** container platforms.

Docker works great for deploying an application inside a server environment. *Docker Hub* maintains a wide variety of Julia images (https://hub.docker.com/_/julia). You can start from such an image, add the apps and files you need, and then deploy the modified image. DockerSandbox.jl (https://juliacontainerization.github.io/DockerSandbox.jl/dev/) lets you run Julia Cmd objects inside a Docker container. There is also a Docker.jl package, which is a Julia interface for the Docker Remote API.

Kubernetes is another alternative. It is an open source orchestration system for the deployment and management of containerized applications. The Kuber.jl package (https://github.com/JuliaComputing/Kuber.jl) is a Kubernetes client, which makes it easy to access Kubernetes clusters from Julia. Dash apps can also be deployed to Kubernetes clusters.

Now that we have explored the possibilities for deploying a Julia app in the cloud, let's discover which Julia web frameworks exist to develop web apps from start to finish.

Choosing the right Julia framework for a project

While it is, in principle, possible to code a web application without using a framework, you don't want to reinvent the wheel(s). Using a framework will save you enormous amounts of time and effort by providing reusable and scalable code for a basic web infrastructure. This way, as a developer, you can concentrate on higher-level application logic.

In this section, we'll review the existing Julia web frameworks and discuss their use cases including their pros and cons. We'll leave out *Dash* and *Franklin*, which are production-ready, and were discussed in previous sections. We can also make a distinction between the following:

- Using Julia for static websites
- A **full stack** framework, which provides all aspects from the data layer to deployment
- A **micro** framework, which only provides the routing part

If you're looking for a simple CGI or FastCGI server, take a look at the `Jaylle` repo at `https://github.com/Jaylle`.

Genie

Genie is the best-known framework for general web development in the Julia ecosystem. It's a full-stack MVC framework and offers an elegant workflow for building web apps. It has a low-code UI layer called `Stipple` and an **object-relational mapper** (**ORM**) layer called `Searchlight`. We'll examine it in much more detail in the coming chapters, starting from *Chapter 4, Building an MVC App Using Genie*.

Bukdu

Bukdu is a full-stack framework that is influenced by the MVC `Phoenix` framework (which is the most popular *Elixir* web framework) as its model. The source repo can be found at `https://github.com/wookay/Bukdu.jl`. There is a Heroku demo at `https://sevenstars.herokuapp.com/`.

Here is some code so that you can get an initial impression (see `Chapter3\using_bukdu1.jl`):

```
# add Bukdu      # in pkg mode
using Bukdu


struct WelcomeController <: ApplicationController      # 1
  conn::Conn
end
```

```
struct RestController <: ApplicationController
  conn::Conn
end


function index(c::WelcomeController)                      # 2
  render(JSON, "Hello World from Bukdu!")
end


function init(c::RestController)
  render(JSON, (:init, c.params.region, c.params.site_id,
    c.params.channel_id))
end


routes() do
  get("/", WelcomeController, index)                      #3A
  get("/init/region/:region/site/:site_id/channel/:channel_
  id/", RestController, init, :site_id => Int, :channel_
  id=>Int)                # 3B
end
Bukdu.start(8080)                                         # 4
# Bukdu Listening on 127.0.0.1:8080
# Task (runnable) @0x00000000735447d0
# INFO: GET    WelcomeController    index              200 /
```

In the preceding code, you can observe the following:

- The Bukdu web server is started on line 4.

- A GET request for the root route is handled on line 3A; it specifies a controller (as defined on line 1) and a function (or index, as defined on line 2) that must be executed. Then, a render function is called that outputs a JSON string.

  Visit http://127.0.0.1:8080 to see the JSON text, Hello World from Bukdu!, rendered in your web browser. Additionally, you can test it in a terminal by invoking curl: curl http://localhost:8080/.

- Line 3B shows more complex route handling where some parameter values, such as :site_id, are taken from the route and displayed in the init function. Here, the following curl command is given: curl localhost:8080/init/region/west/site/1/channel/2/.

The output is returned as follows:

```
["init","west",1,2]
```

In the preceding output, you can see that the :region symbol gets the "west" value, :site_id became 1, and :channel_id became 2.

Bukdu might be a good choice if you love the Phoenix framework. On the other hand, it doesn't seem to be very actively maintained.

Interact

Interact is built upon WebIO, and with it, you can make little GUI apps with web technologies. These apps can be run in a browser, in an *Electron* window (with *Blink*), in a *Jupyter* notebook, or they can work together with *Mux* (see the next section). The package contains UI widgets such as textboxes, checkboxes, buttons, and sliders.

Here is some sample code to display an Electron window showing a drop-down list:

```
# add Interact    # pkg mode
using Interact, Blink

ddn = dropdown(["Getting groceries",
                "Visiting my therapist",
                "Getting a haircut",
                "Paying the energy bill",
                "Blog about workspace management"])
w = Window()
ui = dom"div"(ddn)
body!(w, ui)
```

Interact is documented at https://juliagizmos.github.io/Interact.jl/latest/ and in its GitHub repo at https://github.com/JuliaGizmos/Interact.jl.

Mux

Mux is a micro web framework, particularly useful for apps that deliver endpoint services. Its repo is situated at https://github.com/JuliaWeb/Mux.jl. Mux works nicely together with WebIO and Interact. Here is the example from the preceding section where a widget generated by Interact is served by the Mux web server:

```
using Interact, Mux
ddn = dropdown(["Getting groceries",
```

```
                    "Visiting my therapist",
                    "Getting a haircut",
                    "Paying the energy bill",
                    "Blog about workspace management"])
ui = dom"div"(ddn)
WebIO.webio_serve(page("/", req -> ui), 8080)    # 1
```

In the preceding code, you can observe the following:

- Line *1* starts a web server listening on port 8080, which will return the defined ui in a root request

- The following is the output in a browser:

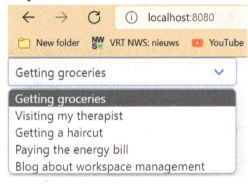

Figure 3.8 – Mux in action

The @app macro lets you define a web server with three routes in just a few lines (see use_mux2.jl):

```
using Mux
@app test = (                                       # 1
  Mux.defaults,
  page(respond("<h1>Hello World from Mux!</h1>")),
  page("/about", respond("<h1>About Mux</h1>")),    # 2
  page("/user/:user", req -> "<h1>Hello,
        $(req[:params][:user])!</h1>"),
  Mux.notfound())
serve(test)                                         # 3
```

In the preceding code, line *1* starts the macro body, and line *3* starts the web server.

Try out the last route by entering http://localhost:8000/user/John in your browser to see the Hello John! message.

Mux is designed in a smart way that enables you to write simple APIs in a couple of lines. The important concept of Mux is middleware: a way to take a request and modify it. An app is just an endpoint function that takes a request and returns a response. You can chain any number of middleware steps with the mux function, as follows:

```
mux(auth, catch_errors, app)
```

Here, auth and catch_errors are middleware functions.

If you want to see some inspiring scripts combining *WebIO* and *Mux*, look at https://github. com/jlgw/julia-scripts. I hope this package gets the attention it deserves to further complete its functionality.

Merly

Merly is a microframework for web development, capable of route declaration and request handling. Its main design purpose is to make development easy, but at the same time, it has been optimized for execution speed. It supports all HTTP methods, including CORS handling and returning configured headers, working with static files, and parsing the page body's content.

The following is a simple example demonstrating Merly's usage:

```
# add Merly    # pkg mode
using Merly


@page "/" HTTP.Response(200,"Hello World from
Merly!")                        # 1
@page "/hello/:user" HTTP.Response(200,string("<b>Hello
",request.params["user"],"!</b>")) # 2


@route POST "/post" HTTP.Response(200,"I did
something!")                        # 3
start(host = "127.0.0.1", port = 8086, verbose = true)    # 4
```

In the preceding code, you can observe the following:

- The @page macro used on line *1* is probably the simplest route-handling mechanism among all Julia frameworks.

- Line *2* takes a parameter user in the /hello path and displays it in the response.

- The @route macro on line *3* lets you specify the HTTP verb to use. Use start() to begin executing the web server (see line *4*).

Start Merly up on a terminal with the following command:

```
julia using_merly.jl
```

Request handling can also be performed with dedicated functions such as `Get`, `Post`, `Put`, and `Connect`.

`Merly` has its own website at `https://merly.vercel.app/` and is fully documented at `https://neomatrixcode.gitbook.io/merly/`. Additionally, the website features different examples of deployment to cloud providers (such as *Heroku*, *Docker*, *Docker Compose*, *Google Cloud*, and *Digital Ocean*).

Dance

The **Dance** framework aims to make it easier for the *data scientist* to present data on the web. To that end, the framework has special methods for converting bidirectionally Julia `Dicts` and `DataFrames` into JSON and JS strings. This enables using existing JS frameworks as a frontend to display data.

Dance has a project generator method called `start_project`(ToDo). This creates a ToDo folder with a startup script, `dance.jl`, and a project-specific script in `src\Todo.jl`. Routes are handled via the `routes.jl` script and can be described through regex patterns, for example, `route(r"/post/dict/(?<value>\d{3})", post_2)`. Additionally, Dance can serve static files.

After having specified `server_host` and `server_port` in `Configuration.jl` as `localhost` and `8080`, respectively, start up the web server using the following command:

```
julia dance.jl
```

Then, opening up a browser with the `localhost:8080` URL shows the `"Hello World"` text.

The integration of `Webpack` and a `NodeJS` environment inside a `Dance.jl` project is made easy through the `DanceWebpack` plugin.

More info about the possibilities of using Dance can be found in the repo at `https://github.com/DanceJL/Dance.jl`.

Summary

In this chapter, we delved into what the Julia community has built over the past 5 years in the way of tools to build web-based applications.

From now on, whenever you need to build a static website, Franklin will be available to you as a valuable choice. If you need to build a cross-platform Electron-based GUI app, you can combine WebIO, Blink, and Interact to accomplish this. If your next project is to build a web dashboard, look no further than Dash for Julia.

Also, we discussed the architecture of microservices, the current state of WASM in the Julia ecosystem, and which cloud platforms are readily available for Julia.

Finally, you got an overview of all mature web frameworks in Julia. You should have a better idea of which to apply when, and you have also gained a head start for working with any of them.

This concludes our overall survey of Julia and other web technologies. In the remaining part of the book, you'll be guided, step by step, on how to build up different sorts of web apps using the Genie web framework.

Part 2: Using the Genie Rapid Web Development Framework

Here, we explore in depth the Genie web framework, continuing our to-do app example. First, we build a complete web app based on the MVC pattern, including unit testing. Then, we enhance this by creating a REST API to expose the data from our application, while writing specific tests and documenting the API.

Deploying our app in a production environment is the next big step. We discuss how we can configure different app environments, how we can use Git and GitHub to establish continuous integration, and how we can improve performance by using PackageCompiler. Then, we learn how to deploy our app in a Docker container and on the Heroku cloud platform.

Lastly, we build a variant of our to-do app in the form of an interactive dashboard that lets us interrogate our data with charts and filters. We also learn how to create mock data for larger-scale testing.

This part of the book comprises the following chapters:

- *Chapter 4, Building an MVC ToDo App*
- *Chapter 5, Adding a REST API*
- *Chapter 6, Deploying Genie Apps in Production*
- *Chapter 7, Adding Authentication to Our App*
- *Chapter 8, Developing Interactive Data Dashboards with Genie*

Building an MVC ToDo App

In this chapter, we will start using the Genie framework (https://genieframework.com/) to build an MVC app. In an app designed according to this pattern, browser requests are sent (routed) to a controller. The controller chooses the view, which is the response displayed on the user's screen (in the browser or otherwise). Additionally, the controller either asks the model for data that is needed in the view or sends new data to the model. Finally, the model is responsible for interacting with the **database (DB)** backend, to retrieve or persist data.

With Genie we can create a variety of applications, including simple server scripts, web services and APIs, and full MVC applications that are suitable for building large, scalable, full-stack web apps. The full-stack MVC apps make use of most of the Genie features and to see them in actions we'll build such an MVC web app throughout this chapter.

Right now, we'll start tackling the Todo app, which was announced in the previous chapters on numerous occasions. This app will allow you to create, modify, and delete to-do items, mark them as completed, and see lists with totals of completed and active to-do items.

Here are the topics we are going to discuss in this chapter:

- Creating and configuring the app
- Setting up the DB
- Creating controllers and views
- Routing requests and rendering various pages
- Testing Genie apps

Technical requirements

We will need to install Genie before we start working through the examples in this chapter. As you already know, installing this package is a breeze.

Start Julia, go into package mode with], and then type in `add Genie`.

The versions of Julia and the packages used in this chapter are *Julia 1.8*, *Genie v5.11*, *SearchLight v2.10*, and *SearchLightSQLite v2.2*. Make sure you use the same or a higher version. You can update a lower version by running `pkg> update Genie`. You can verify the version used by the app by executing `pkg> status` and identifying the entry corresponding to the Genie package.

The source code for this chapter can be found at `https://github.com/PacktPublishing/Web-Development-with-Julia-and-Genie/tree/main/Chapter4`.

Creating and configuring the app

Creating a Genie app is simply done from within the Julia REPL environment. Genie includes various handy generators for bootstrapping new applications. These generators set up the necessary packages and application files in order to streamline the creation of various types of projects, including full-stack (MVC) apps and web services and APIs.

As we're creating an MVC app - which we will name `TodoMVC`, - we'll use the MVC generator. We invoke the `newapp_mvc` generator, passing the name of our new application. Open up the REPL, and at the Julia prompt, type in:

```
using Genie
Genie.Generator.newapp_mvc("TodoMVC")
```

The preceding command creates a folder structure inside the `TodoMVC` root folder of this application. Then, the following dialog is displayed:

```
Please choose the DB backend you want to use:
1. SQLite
2. MySQL
3. PostgreSQL
4. Skip installing DB support at this time

Input 1, 2, 3 or 4 and press ENTER to confirm.
If you are not sure what to pick, choose 1 (SQLite). It is the simplest option to get you started right away.
You can add support for additional databases anytime later.
```

Figure 4.1 – Choosing the DB backend

As MVC apps routinely use DB backends, the generator now gives us the possibility to configure the DB connection. Genie integrates with various DB backends through *SearchLight*, an **object-relational mapper (ORM)** library that provides an API to work with relational DBs. SearchLight makes it very easy to write code that is portable between the supported backends. Our plan is to use SQLite during development for ease of configuring and portability, and PostgreSQL or MySQL/MariaDB in production for high performance and scalability under live online traffic.

In the REPL, after typing `1` and pressing *Enter*, *SQLite* will be installed and configured, together with the *SearchLightSQLite adapter*. You will see that the new *TodoMVC* app has been precompiled.

You will know all has gone well when the following screen appears:

```
[ Info: Precompiling TodoMVC [34ba3b87-ecb2-4d37-9efe-7f04af6730ef]
```

```
| Website  https://genieframework.com
| GitHub   https://github.com/genieframework
| Docs     https://genieframework.com/docs
| Discord  https://discord.com/invite/9zyZbD6J7H
| Twitter  https://twitter.com/essenciary

Active env: DEV


Ready!

[ Info: 2022-05-31 16:52:24
└ Web Server starting at http://127.0.0.1:8000
```

Figure 4.2 – Screen showing the application has started up

The message on the preceding screen implies that Genie has loaded our newly created app and has started a web server on port **8000**. We can check that everything works by opening up a browser with the http://127.0.0.1:8000 URL.

We then see Genie's welcome page:

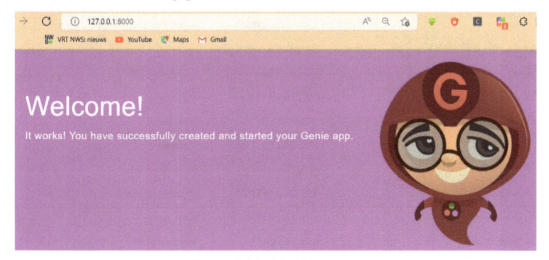

Figure 4.3 – Genie's welcome page

Where does the preceding page come from? Which page is sent to the browser is decided in the `routes.jl` file in the root folder of our app. This file initially has the following contents:

```
using Genie.Router

route("/") do
  serve_static_file("welcome.html")
end
```

The preceding code tells us that for the `http://127.0.0.1:8000/` URL, which is a request for the root (/) path of the web server, the `welcome.html` HTML page from the `public` folder will be shown. As customary in web apps, the `public` folder contains all kinds of web resources such as static HTML pages, CSS, images, fonts, and JS scripts.

On your disk, you can inspect the app's folder structure, as shown in the following screenshot:

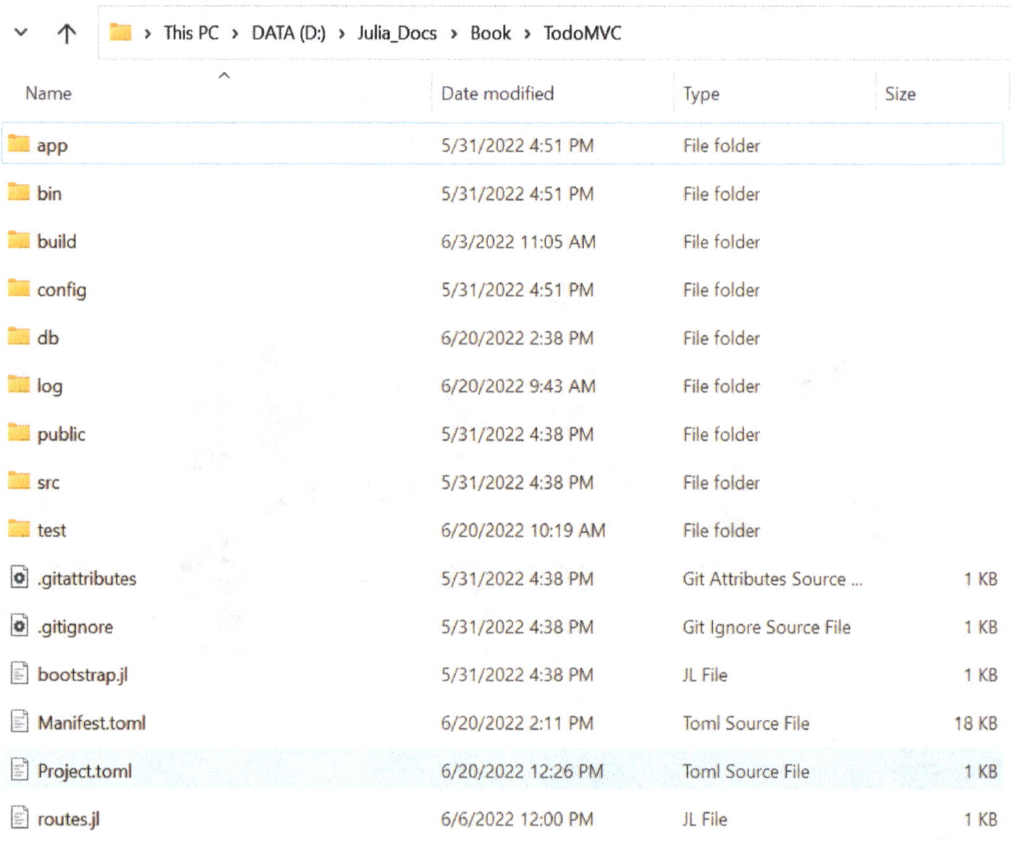
This PC › DATA (D:) › Julia_Docs › Book › TodoMVC
Name
Date modified
Type
Size
app 5/31/2022 4:51 PM File folder
bin 5/31/2022 4:51 PM File folder
build 6/3/2022 11:05 AM File folder
config 5/31/2022 4:51 PM File folder
db 6/20/2022 2:38 PM File folder
log 6/20/2022 9:43 AM File folder
public 5/31/2022 4:38 PM File folder
src 5/31/2022 4:38 PM File folder
test 6/20/2022 10:19 AM File folder
.gitattributes 5/31/2022 4:38 PM Git Attributes Source ... 1 KB
.gitignore 5/31/2022 4:38 PM Git Ignore Source File 1 KB
bootstrap.jl 5/31/2022 4:38 PM JL File 1 KB
Manifest.toml 6/20/2022 2:11 PM Toml Source File 18 KB
Project.toml 6/20/2022 12:26 PM Toml Source File 1 KB
routes.jl 6/6/2022 12:00 PM JL File 1 KB

Figure 4.4 – The root folder of the app

The app folder is the place where you will be working the most: it contains Julia code for the models, views, and controllers.

Congratulations, you just created and started your first Genie app!

Pausing development

Keeping the REPL open isn't always possible. How do you proceed with restarting your work on the app? To do this, just press *Ctrl + C* to stop the web server and quit Julia.

The next time you want to continue work on your web app, go through the following steps:

1. Go to the `TodoMVC` root folder (you can do this from the OS or inside the REPL with the `cd("/path/to/TodoMVC")` function). Verify with `pwd()`.

2. Inside the REPL, in `pkg` mode, type `activate .`:

   ```
   (Book) pkg> activate .
   Activating project at `D:\Julia_Docs\Book\TodoMVC`
   ```

   This activates the local package environment.

3. Back at the Julia prompt, type in the following:

   ```
   julia> using Genie
   julia> Genie.loadapp()
   ```

   This loads the app's environment.

   You can do the same from a Terminal session while in the app's root folder by executing `bin/repl`.

Then, start the web server by typing in `up()`, and continue developing!

Launching the `bin/server` script does all of the preceding actions, starting the web server in a non-interactive mode, while not needing access to the REPL. On Windows, use the same command, replacing / with \.

Setting up the DB

Various relational DB backends support different features and flavors of SQL. Using *SearchLight* ensures that our DB-interaction code can be ported across the different supported backends. This pattern also covers table creation and modification, which is done via *migration* scripts. Besides being DB agnostic, migration scripts provide another very important advantage: they allow versioning and automating/repeating table creation and modification operations, for example, between multiple team members or when deploying the app in production.

Before we can use migrations to create our DB table(s), we need to set up the migration infrastructure. This is a table stored in the app's DB, where SearchLight keeps track of the various migration scripts. This is easily done with another generator. Type in the following at the Julia prompt:

```
julia> using SearchLight
julia> SearchLight.Migrations.init()
```

Then, we will get the following output:

```
┌ Info: CREATE TABLE `schema_migrations` (
│           `version` varchar(30) NOT NULL DEFAULT '',
│           PRIMARY KEY (`version`)
└       )
[ Info: Created table schema_migrations
```

Creating our table

Our application will need a DB table to store the todos; let's call it Todo. Also, we need a way to interact with this DB table, in order to store, retrieve, update, and delete the to-do items. This is done using *models* – the *M* in the *MVC* stack.

SearchLight has a series of generators that allow us to quickly create models and their respective migrations, plus a few other useful files. Let's create all of this with the following command:

```
julia> SearchLight.Generator.newresource("Todo")
```

Genie displays the following output to signal that four Julia files have been created:

```
[ Info: New model created at TodoMVC/app/resources/todos/Todos.
jl
[ Info: New table migration created at TodoMVC/db/
migrations/<timestamp>_create_table_todos.jl
[ Info: New validator created at TodoMVC/app/resources/todos/
TodosValidator.jl
[ Info: New unit test created at TodoMVC/test/todos_test.jl
```

A resource represents a business entity or a piece of data (in our case, a to-do item) implemented in code through a bundle of files serving various roles. For now, we'll focus on the model and the migration – but notice that SearchLight has also created a validator and a test file. We'll get back to these later in the *Adding validation* and *Running tests with Genie* sections.

As you can see in the preceding output, the migration file has been created inside the db/migrations/ folder. The filename ends in _create_table_todos.jl and begins with a timestamp of the moment the file was created, which is something like 2022060109133379_create_table_ todos.jl. The purpose of timestamping the migration file is to reduce the risk of name conflicts when working with a team – but also to inform SearchLight about the creation and execution order of the migration files.

Let's check out the migration file. It looks like this:

```
module CreateTableTodos
import SearchLight.Migrations: create_table, column, columns,
pk, add_index, drop_table, add_indices
function up()
  create_table(:todos) do
    [
      pk()
      column(:column_name, :column_type)
      columns([
        :column_name => :column_type
      ])
    ]
  end
  add_index(:todos, :column_name)
  add_indices(:todos, :column_name_1, :column_name_2)
end

function down()
  drop_table(:todos)
end
end
```

In the preceding code, we can see that SearchLight has added some boilerplate code to get us started – we just need to fill up the placeholders with the names and properties of our table's columns. The Migrations API should be pretty self-explanatory, but let's go over it quickly. We have two up and down functions. In migrations parlance, the up function is used to apply the DB modification logic. Any changes we want to make should go into the up function. Conversely, the down function contains logic for undoing the changes introduced by up.

Moving on to the contents of the up function, it creates a table called todos with the create_table(:todos) command, adds a primary key (pk()), and then provides a boilerplate for adding a number of columns and indices. The down function deletes the table with the drop_table(:todos) command, undoing the effects of up.

In the spirit of traditional TodoMVC apps, we'll keep it simple, and we'll only store the to-do item itself (we'll call it todo) and whether or not it has been completed (when first created, a to-do item is not complete). Let's set up the up logic, only showing what has changed in the preceding code:

```
function up()
  create_table(:todos) do
    [
      pk()
      column(:todo, :string)
      column(:completed, :bool; default = false)
    ]
  end

  add_index(:todos, :completed)
end
```

Now, we're ready to do our migration, in this case, executing the code within the up function. The SearchLight.Migration API provides a series of utilities to work with migrations, for instance, to keep track of which migrations have been executed and to execute migrations in the correct order. We can check the status of our migrations as follows:

```
julia> SearchLight.Migrations.status()
```

The output will be as follows:

```
[ Info: SELECT version FROM schema_migrations ORDER BY version
DESC
|   | Module name & status                          |
|   | File name                                     |
|---|-----------------------------------------------|
|   |                      CreateTableTodos: DOWN   |
| 1 |          <timestamp>_create_table_todos.jl    |
```

As expected, our migration is down, meaning that we haven't run the up function to apply the changes to the DB. Let's do it now:

```
julia> SearchLight.Migrations.up()
```

We can see all of the steps executed by the up function:

```
[ Info: SELECT version FROM schema_migrations ORDER BY version
DESC
[ Info: CREATE TABLE todos (id INTEGER PRIMARY KEY , todo
TEXT  , completed BOOLEAN  DEFAULT false  )
[ Info: CREATE  INDEX todos__idx_completed ON todos (completed)
[ Info: INSERT INTO schema_migrations VALUES
('2022052910095674')
[ Info: Executed migration CreateTableTodos up
```

Now check with the preceding `status()` command that the status is now up.

Setting up the model

SearchLight generates a multitude of SQL queries that are optimized for the configured DB backend (in our case, this is SQLite). This idea is taken to its next step by *models*. Models are even more powerful constructs compared to migrations. Migrations manipulate the table structure, while models allow us to manipulate the data. A model is a Julia struct whose fields (properties) map to the table columns that we want to control. By setting up these structs, we retrieve data from our DB tables – and by changing the values of their fields, we write data back to the DB.

Remember that our model was created in the `app/resources/todos/` folder, under the name of `Todos.jl`. Let's open it in our editor:

```
module Todos
import SearchLight: AbstractModel, DbId
import Base: @kwdef
export Todo


@kwdef mutable struct Todo <: AbstractModel
  id::DbId = DbId()
end


end
```

Similar to the migration, SearchLight has set up a good amount of boilerplate to get us started. The model struct is included in a module. Notice that the name of the Todos module is pluralized, like the name of the todos table, while the Todo struct is singular. The table contains multiple todos; and each Todo struct represents one row in the table, that is, one to-do item.

The struct already includes the id field corresponding to the primary key. Let's add two other fields corresponding to the todo description and the completed status. These fields must match the names of the types we declared in the migration, so we change our Todo struct definition to the following:

```
@kwdef mutable struct Todo <: AbstractModel
    id::DbId = DbId()
    todo::String = ""
    completed::Bool = false
end
```

Let's give our model a try:

```
julia> using Todos
```

The preceding command precompiles the Todos model.

We can ask SearchLight to find all the todos with the all command:

```
julia> all(Todo)
[ Info: 2022-05-29 12:58:03 SELECT "todos"."id" AS "todos_id",
"todos"."todo" AS "todos_todo", "todos"."completed" AS "todos_
completed" FROM "todos" ORDER BY todos.id ASC
Todo[]
```

The output shows the generated SQL command, along with an empty Todo[] vector, because we haven't yet added any to-do items. It's time to create our first to-do item:

```
julia> my_first_todo = Todo()
```

We've just created our first to-do item, as you can see in the following output:

```
Todo
| KEY             | VALUE |
|-----------------|-------|
| completed::Bool | false |
| id::DbId        | NULL  |
| todo::String    |       |
```

However, this contains no data, so it is not very useful. We should store something useful in it:

```
julia> my_first_todo.todo = "Build the Genie TodoMVC app"
```

Now, to store what you have added, run the following:

```
julia> save!(my_first_todo)
```

This generates the following output, showing the INSERT and SELECT SQL, and the contents of the only record so far in the todos table:

```
[ Info: INSERT  INTO todos ("todo", "completed") VALUES ('Build
the Genie TodoMVC app', false)
[ Info: SELECT CASE WHEN last_insert_rowid() = 0 THEN -1 ELSE
last_insert_rowid() END AS LAST_INSERT_ID
[ Info: SELECT "todos"."id" AS "todos_id", "todos"."todo" AS
"todos_todo", "todos"."completed" AS "todos_completed" FROM
"todos" WHERE "id" = 1 ORDER BY todos.id ASC


Todo
| KEY             | VALUE                        |
|-----------------|------------------------------|
| completed::Bool | false                        |
| id::DbId        | 1                            |
| todo::String    | Build the Genie TodoMVC app  |
```

The save! function will persist the to-do data to the DB, modifying our to-do instance by setting its id field to the row ID that was retrieved from the DB operation. If the DB operation fails, an exception is thrown.

SearchLight is smart and runs the correct queries, depending on the context. In the preceding example, it generated an INSERT query to add a new row – but when changing an instance that already has data loaded from the DB, it will generate an UPDATE query instead. Try this out by entering the save!(my_first_todo) command once more. Also, add some more to-do records to our table, update some of them, and read them back out.

Now we are done setting up the DB interaction layer (the model layer). Next, we'll discuss the view and the controller layers of our Genie Todo MVC application.

Creating controllers and views

In MVC applications, the views display the data, which comes from the model layer to the user. However, the views do not interact directly with the model layer. Instead, they interact with the controller layer. The controller layer is responsible for handling user input and updating the model data, too. Every time a web request is made to the server, first, the controller is invoked, reading and/or modifying the model data. Then, this model data is passed to the view layer, which formats and displays the data to the user. Let's see this in action in our app. Genie's generator will create a controller for us:

```
julia> Genie.Generator.newcontroller("Todo")
```

The controller file is in the same location as our model, as indicated by the following output:

```
[ Info: New controller created at TodoMVC/app/resources/todos/
TodosController.jl
```

Let's add logic to display all the todos. We'll start by adding a function (let's call it index), which retrieves all the todos from the DB and renders them to the user. Here is our modified TodosController. jl file:

```
module TodosController

using TodoMVC.Todos
using Genie.Renderers, Genie.Renderers.Html

function index()
  html(:todos, :index; todos = all(Todo))
end
end
```

It's as simple as this: we retrieve all the to-do items using the all function from SearchLight, and pass them to the index view, within the todos resource folder.

Now it's time to add a simple view file – create the app/resources/todos/views folder and create an index.jl.html file. We can do this from within the REPL:

```
julia> mkdir("app/resources/todos/views")
julia> touch("app/resources/todos/views/index.jl.html")
```

Genie supports a variety of languages for views, including pure *Julia*, *Markdown*, and *HTML* with embedded Julia. Any file that has .jl.html as an extension is a *view template*, which is like a screen layout where data will be filled in by our app. Our index.jl.html file will be mostly written with HTML, and we'll use Julia language constructs (such as if, for, and more) and Julia variable interpolation to make the output dynamic.

Now, edit the index.jl.html file and add the following code:

```
<% if isempty(todos) %>
  <p>Nothing to do!</p>
<% else %>
  <ul>
    <% for_each(todos) do todo %>
      <li>
        <input type="checkbox" checked="$(todo.completed ? true
          : false)" />
        <label>$(todo.todo)</label>
      </li>
    <% end %>
  </ul>
<% end %>
```

In the preceding HTML code, we use a series of Julia language constructs to dynamically generate the HTML. The if statement checks whether the todos vector is empty and, if so, displays a message to the user. Otherwise, it iterates over the todos vector and displays each to-do item. Julia code blocks are delimited by <% %> tags, while for outputting values, we resort to the $(...) syntax for string interpolation. Also, notice the use of the for_each function – this is a helper provided by Genie to iterate over a collection and automatically concatenate the output of the loop and render it into the view.

We're almost ready to view our todos on the web. But there is one thing missing: we need to register a *route*: a mapping between a URL that will be requested by the users and the controller function that will return the response. That's what we'll do next.

Routing requests and rendering various pages

Let's add a route to our app. Edit the routes.jl file inside the TodoMVC folder and change it to look like this:

```
using Genie
using TodoMVC.TodosController

route("/", TodosController.index)
```

The route (...) line tells us that when the root folder of the app is requested, the index function in TodosController.jl must be called. From the previous section, we already know what will then be displayed. We can access our todos at http://localhost:8000/, and we can now see the one to-do item we previously created:

- ☑ Build the Genie TodoMVC app

Figure 4.5 – The index page

Using a layout file

When rendering a view file, by default, it is automatically wrapped by a layout file. The role of the layout is to render generic UI elements that are present on multiple pages, such as the main navigation or the footer. The default layout file is located in the app/layouts folder and is called app.jl.html. Let's use it to style our todos a bit. Edit the app.jl.html file and make it look like this:

```
<!DOCTYPE html>
<html lang="en">
  <head>
    <meta charset="utf-8" />
    <title>Genie Todo MVC</title>
    <link href="https://cdn.jsdelivr.net/npm/bootstrap@5.0.2/
      dist/css/bootstrap.min.css" rel="stylesheet">
  </head>
  <body>
    <div class="container">
      <h1>What needs to be done?</h1>
      <%
        @yield
      %>
    </div>
  </body>
</html>
```

Arguably, one of the most important elements of the layout file is the @yield macro. This macro is used to render the content of the view file. In the preceding example, the @yield macro is used to render the contents of the index.jl.html file.

We have included the *Bootstrap CSS* library in our app.jl.html file. This library provides a lot of useful styles and components for our Todo app, and we'll use some of them in our view. We've added a div with the class container to make our layout responsive and centered. Also, we have an h1 element to display the title of our app. Next, make sure that the index.jl.html file is updated, as follows, by adding some classes from Bootstrap:

```
<% if isempty(todos) %>
  <p>Nothing to do!</p>
<% else %>
  <div class="row">
    <ul class="list-group">
      <% for_each(todos) do todo %>
        <li class="list-group-item form-check form-switch">
          <input type="checkbox" checked="$(todo.completed
          ? true : false)" class="form-check-input"
          id="todo_$(todo.id)"  value="$(todo.id)" />
          <label class="form-check-label"
          for="todo_$(todo.id)">$(todo.todo)</label>
        </li>
      <% end %>
    </ul>
  </div>
<% end %>
```

Our to-do list looks much better with this Bootstrap checkbox class already, as you can see in the following screenshot:

What needs to be done?

Build the Genie TodoMVC app

Figure 4.6 – The bootstrapped index page

In order to build more complex views, Genie also provides another feature, called view partials. Let's see what they are.

View partials

View partials are small pieces of code that can be reused in multiple views. Typically, their name starts with _. Let's add a view partial to our app that contains a form for creating new todos. We'll create this file in the app/resources/todos/views folder and name it _form.jl.html. You can do this from the REPL by issuing the following command:

```
julia> touch("app/resources/todos/views/_form.jl.html")
```

Then, add the following content to the _form.jl.html file:

```
<div class="row">
  <form method="POST" action="/todos">
    <div class="input-group mb-3">
      <input type="text" class="form-control"
      placeholder="Add a new todo">
      <input type="submit"
        class="btn btn-outline-secondary" value="Add">
    </div>
  </form>
</div>
```

In the preceding code, we have added a form with one text field for entering the new to-do item and a submit button. In order to include the partial into our view, we'll use the partial function. Add the following code to the end of the index.jl.html file:

```
<% partial("app/resources/todos/views/_form.jl.html") %>
```

In order for our form to work as expected, we need to add the corresponding route and the controller function. To add the route, edit the routes.jl file and add the following code:

```
route("/todos", TodosController.create, method = POST)
```

Notice that the /todos route corresponds with the action attribute of the form.

For the controller function, edit the TodosController.jl file and add the following code:

```
using Genie.Router
using SearchLight

function create()
  todo = Todo(todo = params(:todo))            # 1
```

```
  if save(todo)                          # 2
    redirect("/?success=Todo created")   # 3
  else
    redirect("/?error=Could not save
              todo&todo=$(params(:todo))")
  end
end
```

In the preceding code, after adding a few extra `using` statements that give us access to the `redirect` and `save` methods, we create a new `Todo` instance on line *1*, using data we retrieve through `params`. This `params` function is a mapping to the `params` struct in the Genie `Router` module, which contains the request variables collection. Then, on line 2, we save the `Todo` instance to the DB. If the `save` operation succeeds, we `redirect` the user to the index page with a success message on line *3*. Otherwise, in the `else` branch, we redirect the user to the index page with an `error` message and the current to-do item, to fill up the new to-do field with the to-do's description.

Now, let's add the necessary extra code to the frontend. First, to handle success and error messages in the `index.jl.html` file, let's add another view partial to handle the messages. Add the following on the very first line of `index.jl.html`:

```
<% partial("app/resources/todos/views/_messages.jl.html") %>
```

Now, create this file in the REPL with `touch("app/resources/todos/views/_messages.jl.html")` and edit it as follows:

```
<% if ! isempty(params(:success, "")) %>
  <div class="alert alert-success" role="alert">
    <% params(:success) %>
  </div>
<% elseif ! isempty(params(:error, "")) %>
  <div class="alert alert-danger" role="alert">
    <% params(:error) %>
  </div>
<% else %>
  <br/>
<% end %>
```

In the _messages.jl.html partial, we are checking whether there is a :success parameter in the query string. If there is, we display a success message. Otherwise, we check whether there is a :error parameter in the query string. If there is, we display an error message. Otherwise, we display nothing.

Finally, in the _form.jl.html file, we need to update the input tag to automatically display the to-do item that the user entered. Replace the line where the text input tag is defined with the following code (we've added the name and value attributes at the end):

```
<input type="text" class="form-control" placeholder="Add a new
todo" name="todo" value='$(params(:todo, ""))' />
```

The to-do name attribute links to params(:todo) on line *1* of the create function. Notice that we're using single quotes (') for the value attribute since we're using double quotes inside it.

Now we can add the to-do items in our app:

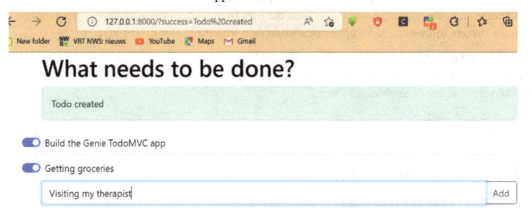

Figure 4.7 – Adding the to-do items

So far, everything looks great, but there is a problem. Our application allows users to create new todos, but they can also create empty todos – which are not very useful. We need to add some validation to our application to prevent users from creating empty todos.

Adding validation

Validations are performed by *model validators*. They represent a collection of validation rules that are applied to a model's data. The TodosValidator.jl file is already included in our application as it was created together with the model. If we open it, we'll see that it includes a few common validation rules, including a not_empty rule:

```
function not_empty(field::Symbol, m::T)::ValidationResult where
{T<:AbstractModel}
```

```
isempty(getfield(m, field)) && return
  ValidationResult(invalid, :not_empty, "should not be
    empty")
ValidationResult(valid)
```
end

All we need to do is to update our Todo model to declare that the todo field should be validated by the not_empty rule. Add the following code to the Todos.jl model file:

```
using SearchLight
using TodoMVC.TodosValidator
import SearchLight.Validation: ModelValidator, ValidationRule

SearchLight.Validation.validator(::Type{Todo}) =
ModelValidator([
  ValidationRule(:todo, TodosValidator.not_empty)
])
```

Now, in the TodosController.jl file, we add a using SearchLight.Validation command and modify the create function to enforce validations. Add the following code after the first line in the function and before the if block:

```
validator = validate(todo)
if haserrors(validator)
  return redirect("/?
    error=$(errors_to_string(validator))")
end
```

Now our application no longer allows the creation of empty to-do items! If you add an empty item, the following error message text appears on the top of the screen:

What needs to be done?

Todo should not be empty

Figure 4.8 – Adding an empty item

Now, let's see how we can implement changing the to-do info and removing entire to-do records (for example, the empty ones).

Updating and deleting todos

The most satisfying part of having a to-do list is marking the items as completed. As it is right now, the application allows us to toggle the completed status of a to-do item, but the change is not persisted to the DB. Let's fix this.

First, let's add a new route and the associated controller function to allow us to toggle the completed status of our to-do items. Add the following code to the `routes.jl` file:

```
route("/todos/:id::Int/toggle", TodosController.toggle, method
= POST)
```

Notice the `:id::Int` component of the route. This is a dynamic route that will contain the `id` of the to-do item that we want to toggle. Also, the route only matches integer values, making sure that incorrect values cannot be passed to the controller function.

Now, for the controller function, edit the `TodosController.jl` file and add the following code:

```
using Genie.Renderers.Json

function toggle()
  todo = findone(Todo, id = params(:id))
  if todo === nothing
    return Router.error(NOT_FOUND, "Todo item with id
      $(params(:id))", MIME"text/html")
  end

  todo.completed = ! todo.completed

  save(todo) && json(:todo => todo)
end
```

In the `toggle` function, we are finding the to-do item with the given `id` through the `findone` function. If the to-do item is not found, we return an error page. Otherwise, we toggle the completed status of the to-do item and save it to the DB, before returning the to-do's data in JSON format.

The reason for returning a JSON response from the `toggle` function is that we want to update the to-do item in the browser without reloading the page. We do this by using *JS* to make an *AJAX* request and then consuming the response data in JS (note that an AJAX request is just an HTTP request made by the JavaScript code to the server). The simplest way to achieve this is by returning a JSON response that can be easily parsed by our JS code. The `json` function is a helper function (available in the `Genie.Renderers.Json` module) that will return a JSON response with the given data. Let's see how we can enhance our app with custom JavaScript!

Enhancing our app with custom JavaScript and CSS

For making the AJAX request, we'll use a JS library called **Axios** (`https://axios-http.com`). First, we'll download the script by adding the following tag to our `app/layouts/app.jl.html` layout file right above the closing `</body>` tag:

```
<script src="https://unpkg.com/axios/dist/axios.min.js"></
script>
```

While we're at it, let's also include **Cash**, a very small JS utility library that makes manipulating the DOM a breeze (`https://github.com/fabiospampinato/cash`). Again, let's load it right above the closing `</body>` tag:

```
<script src="https://cdn.jsdelivr.net/npm/cash-dom@8.1.1/dist/
cash.min.js"></script>
```

Also, we need to create and include an extra JS file where we will put our own code. Any file that we place inside the `public/` folder, in the root of our app, will be available to include in our HTML views. We'll create a new file, called `app.js`, in `public/js/`: `julia> touch("public/js/app.js")`. Then, we'll add it to our layout file (you guessed it – also right before the closing `</body>` tag):

```
<script src="/js/app.js"></script>
```

Now edit the `app.js` file and put this in it:

```
$(function() {
  $('input[type="checkbox"]').on('change', function() {
    if (this.checked) {
      $(this).siblings('label').addClass('completed');
    } else {
      $(this).siblings('label').removeClass('completed');
    }
```

```
    });
})
```

Let's start with the custom CSS, and use it to style our to-do items. We'll add a CSS file to our app:

```
julia> touch("public/css/app.css")
```

Then, insert the following `<link>` right before the closing `</head>` tag in our layout file:

```
<link href="/css/app.css" rel="stylesheet" />
```

Now, add the following CSS rules to the newly created `app.css` file:

```css
.completed {
    text-decoration: line-through;
    color: #d9d9d9;
}
```

Now refresh the page with the to-do list and toggle the checkboxes. You can see how the to-do items are grayed out when they are marked as completed. However, the actual state of the to-do items is not persisted to the DB yet. In order to accomplish this, add the following code snippet to the `app.js` file to perform a POST request via AJAX to the `/todos/:id/toggle` route and update the to-do item's `completed` status:

```javascript
$(function() {
  $('input[type="checkbox"]').on('change', function() {
    axios({
      method: 'post',
      url: '/todos/' + $(this).attr('value') + '/toggle',
      data: {}
    })
    .then(function(response) {
      $('#todo_' + response.data.id.value).first().checked
        = response.data.completed;
    });
  });
});
```

Updating to-do items

Now that we can change the `completed` status of to-do items, we can also allow users to edit the `todo` items themselves. We can do this by adding a double-click event to our to-do items that enable editing mode. Then, we capture the *Enter* key to save the changes (while the *Esc* key will cancel the changes). Add the following code to the `app.js` file to enable this functionality:

```
$(function() {
  $('li > label').on('dblclick', function() {
    $(this).attr('contenteditable', true);
  });
  $('li > label').on('keyup', function(event) {
    if (event.keyCode === 13) {
      $(this).removeAttr('contenteditable');
      axios({
        method: 'post',
        url: '/todos/' + $(this).data('todo-id') +
             '/update',
        data: { todo: $(this).html() }
      })
      .then(function(response) {
        $('label[data-todo-id="' + response.data.id.value +
          '"]').first().html(response.data.todo);
      });
    } else if (event.keyCode === 27) {
      $(this).removeAttr('contenteditable');
      $(this).text($(this).attr('data-original'));
    }
  });
});
```

In order for the JS code to work, we need to make a modification to our `app/resources/todos/views/index.jl.html` view. Replace the line that adds the `<label>` element with the following:

```
<label class='form-check-label $(todo.completed ? "completed"
: "")' data-original="$(todo.todo)" data-todo-id="$(todo.
id)">$(todo.todo)</label>
```

While we're at it, we'll move the partial _form.jl.html file at the top right beneath _messages.jl.html.

Now, let's add a new route and controller function to allow us to update the description of the to-do items. Add the following code to the routes.jl file:

```
route("/todos/:id::Int/update", TodosController.update, method
= POST)
```

Then, in the TodosController.jl file, add the update function:

```
using Genie.Requests
function update()
  todo = findone(Todo, id = params(:id))
  if todo === nothing
    return Router.error(NOT_FOUND, "Todo item with id
      $(params(:id))", MIME"text/html")
  end

  todo.todo = replace(jsonpayload("todo"), "<br>"=>"")
  save(todo) && json(todo)
end
```

You'll notice that the update function is very similar to the toggle function. The only difference is that we are updating the to-do item's description instead of its completed status. The value of the todo parameter is the value of the todo field in the JSON payload, which we access through the jsonpayload function provided by the Genie.Requests module. In addition, we do some basic input cleaning by removing any <br> tags from the description.

Deleting to-do items

It can be useful to also allow the users to remove todos, whether they have been completed or not. We can do this by adding a delete button to each to-do item. Update the index.jl.html view to add the following code right after the <label> tag (above the closing </li> tag):

```
<button class="btn btn-outline-danger invisible" type="button"
value="$(todo.id)">Delete</button>
```

Next, add the following code to the app.js file:

```
$(function() {
  $('li').on('mouseenter', function() {
```

```
    $(this).children('button').removeClass('invisible');
  });
  $('li').on('mouseleave', function() {
    $(this).children('button').addClass('invisible');
  });
  $('li > button').on('click', function() {
    if ( confirm("Are you sure you want to delete this
        todo?") ) {
      axios({
        method: 'post',
        url: '/todos/' + $(this).attr('value') + '/delete',
        data: {}
      })
      .then(function(response) {
        $('#todo_' +
          response.data.id.value).first().parent().remove();
      });
    }
  });
});
```

What have we done so far? Well, we have added a new button to each to-do item that allows the user to delete the to-do item. The button is invisible by default, but when the user hovers over the to-do item, the button becomes visible. When the user clicks on the button, a confirmation dialog is displayed. If the user confirms, an AJAX request is sent to the /todos/:id/delete route to delete the to-do item. The response from the server is then used to remove the to-do item from the page.

Now we add the server-side logic. First, add the following code to the routes.jl file:

```
route("/todos/:id::Int/delete", TodosController.delete, method
= POST)
```

Then, in the TodosController.jl file, add the delete function:

```
function delete()
  todo = findone(Todo, id = params(:id))
  if todo === nothing
    return Router.error(NOT_FOUND, "Todo item with id
      $(params(:id))", MIME"text/html")
```

```
    end

    SearchLight.delete(todo)
    json(Dict(:id => (:value => params(:id))))
  end
```

The `SearchLight.delete` function removes the to-do item from the DB and returns the modified to-do item, setting its `id` value to `nothing` to indicate that the instance is no longer persisted in the DB. However, our frontend needs the to-do item's `id` value to be returned so that it can be removed from the page. We can accomplish this by returning the to-do item's `id` value in the JSON response.

Here is a screenshot of our app where you can see that the completed stat's of the third item has been set to true, and you can also see the **Delete** button when you want to completely delete the record from the DB:

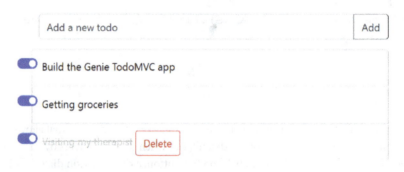

Figure 4.9 – Deleting to-do items

Now you're going to quickly have a lot of to-do items! Wouldn't it be nice if you could select and only count those that have been completed or those that are still to be done?

Aggregate values and filters

The last piece of functionality of our TodoMVC application is to allow the users to filter the to-do items by their `completed` status. We can accomplish this by adding a new toolbar with three buttons, representing the three possible filters for our list: `All`, `Not done`, and `Completed`. For each of the buttons, we also want to show a count of the actual number of todos that match the filter.

To keep our view code clean and easy to maintain, we'll create a new view partial to host our new UI elements. Add the following code to the `index.jl.html` view, right at the bottom:

```
<% partial("app/resources/todos/views/_filters.jl.html") %>
```

Now, create the preceding view partial with the following:

```
julia> touch("app/resources/todos/views/_filters.jl.html")
```

Edit the _filters.jl.html file, and add the following code:

```
<div class="btn-group" role="group">
  <a class="btn btn-outline-primary $(active())" href="/">
    All <span class="badge bg-secondary">$(alltodos)</span>
  </a>
  <a class='btn btn-outline-primary $(active("notdone"))'
      href="/?filter=notdone">
    Not done <span class="badge
      bg-secondary">$(notdonetodos)</span>
  </a>
  <a class='btn btn-outline-primary $(active("done"))'
      href="/?filter=done">
    Completed <span class="badge
      bg-secondary">$(donetodos)</span>
  </a>
</div>
```

Let's unpack this code. We have three <a> elements, styled as buttons, and rendered as a toolbar (thanks to the *Bootstrap* library that we included on our page in the *Using a layout file* section). Within the HTML code, we interpolate a few pieces of Julia code that make our output dynamic. That is, for each button, we invoke a function called active that adds an active CSS class if the button matches the active filter. When the user clicks on a button, a GET request is sent with the :filter parameter set from the URL value of href="/?filter=done". Within each button, inside the nested <span> tag, we substitute the number of todos that matches the filter.

As such, we need to make sure that these values are defined and available in the view layer. We can do this by adding the following code to the TodosController.jl file (update the index function to look like this, and add the extra using statement):

```
using TodoMVC.ViewHelper

function index()
  notdonetodos = count(Todo, completed = false)
  donetodos = count(Todo, completed = true)
  alltodos = notdonetodos + donetodos
```

```
todos = if params(:filter, "") == "done"
  find(Todo, completed = true)
elseif params(:filter, "") == "notdone"
  find(Todo, completed = false)
else
  all(Todo)
end


html(:todos, :index; todos, notdonetodos, donetodos,
  alltodos, ViewHelper.active)
end
```

In the preceding snippet, we use the `count` function to run a count query against the DB. Then, matching the filters from the request params, we use the `find` function to retrieve the todos in each filter. As you might have guessed, the count and find functions are imported from the SearchLight package. Also, we reference a new module, `TodoMVC.ViewHelper`, and pass `ViewHelper.active` into the view together with all the count values that we computed. In an MVC application, helpers are modules that bundle functions that are used in the view layer in order to keep the view code **DRY** (which stands for **Don't Repeat Yourself** and means you must avoid duplicating code in an app) and simple. In order for our code to work, we need to define the new module and the active function.

First, create the helper file *if* it doesn't already exist:

```
julia> touch("app/helpers/ViewHelper.jl")
```

Next, edit it and add the following code:

```
module ViewHelper

using Genie

function active(filter::String = "")
  params(:filter, "") == filter ? "active" : ""
end

end
```

The `active` function is pretty simple. It takes a single optional argument, `filter`. If this matches the current GET request's `filter` parameter, the function returns the `"active"` string, which adds the background color to the button.

Now, try out all the different functionalities that we added in this chapter. Here is a screenshot showing all the records:

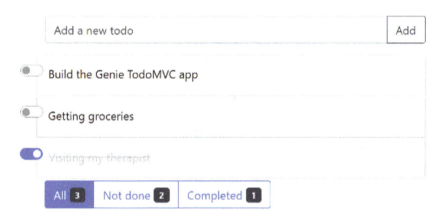

Figure 4.10 – The list of All to-do items

This is a screenshot showing the **Not done** todos:

What needs to be done?

Figure 4.11 – The Not done to-do items

And this is a screenshot showing the **Completed** todos (with their text showing with the `line-through` decoration applied):

Figure. 4.12 – The Completed to-do items

But we know that such manual testing will never be enough. Let's explore how Genie allows you to add automated tests to your app.

Testing Genie apps

Testing is a critical part of developing high-quality software that is easy to scale and maintain. Some developers take this to the extreme and prefer to write the tests before the code is written, starting with failing tests that describe the desired behavior and APIs, and making sure that the tests pass as they implement the minimum necessary feature. This way of working is known as **Test-Driven Development** (**TDD**). TDD is a good concept, but it might not be always possible or efficient. Our standpoint is that it's fine whether the tests are written before or after the actual code it tests. However, when the tests are written after the code is written, the project should not be considered complete until it has proper test coverage and all the tests pass.

The Julia community recognizes the importance of tests, and for this reason, Julia has a built-in testing framework under the `Test` module. In addition, there are multiple packages that improve upon the `Test` API.

Genie uses the testing features available in Julia and some third-party packages to set up a ready-to-use testing environment for Genie applications. Genie handles most of the configuration and boilerplate code for you, so you can focus on writing your tests. So, how does it work?

When a new Genie MVC app is created, a `test/` folder is automatically added. Inside the `test/` folder, Genie creates a new Julia project, with all the necessary files and dependencies to test the Genie app. The main file for running the tests is `runtests.jl`, which is also the standard file for running tests in the Julia ecosystem. The `runtests.jl` file loads the Genie application, making its various MVC parts available for the tests. In addition, Genie adds and configures `TestSetExtensions`, a convenient package that improves upon Julia's default testing capabilities (`https://github.com/ssfrr/TestSetExtensions.jl`) and makes the running of the tests easier and more modular. With `TestSetExtensions`, all we need to do is add test files inside the `test/` folder and they will be automatically executed.

Writing and running our first tests

There are multiple types of tests, some of the most common ones being unit tests and integration tests. Unit tests are tests that test the behavior of a single piece of code. Integration tests are tests that test the behavior of a whole application. When it comes to MVC applications, unit tests are usually focused on testing the models, while integration tests cover larger features that involve at least two layers of the MVC stack.

If you check the `test/` folder, you will see that a test file has already been added, called `todos_test.jl`. This was created automatically by `SearchLight` when we created our model. The file doesn't include any meaningful tests (that's our job!), but we already have more than enough to get us started.

The `todos_test.jl` test file includes the necessary dependencies and defines a test set with a basic test:

```
using Test, SearchLight, Main.UserApp, Main.UserApp.Todos

@testset "Todo unit tests" begin
  ### Your tests here
  @test 1 == 1
end;
```

Let's make sure that everything is working as expected. We can run the tests by executing the `runtests.jl` file in the `test` folder, as follows:

```
julia --project runtests.jl
```

Our tests run successfully, as shown by the following output:

```
todos_test: .

Test Summary: | Pass  Total  Time
TodoMVC tests |    1      1  0.8s
```

But looking at our output, we can see that an exception has been thrown while loading the app's initializers:

```
Loading initializers
ERROR: SearchLight.Exceptions.
MissingDatabaseConfigurationException("DB configuration for
test not found")
...output omitted ...
in expression starting at TodoMVC/config/initializers/
searchlight.jl:13
```

We'll address this exception next.

Configuring the test DB

What Genie is telling us is that we have not configured a test DB for SearchLight. This is a very important point: Genie automatically sets up and runs the application in a test environment to make sure that we don't accidentally run tests against our development or production DBs. However, because our tests don't use the DB yet, none of the tests were impacted. Nevertheless, let's make sure that we have a test DB configured.

Setting up the test DB is straightforward. Edit the db/connection.yml file and add the following lines:

```
test:
  adapter:  SQLite
  database: db/test.sqlite3
```

Now, if we run the tests again, the error has gone. However, the included test is not very useful beyond making sure that our testing environment is working. We need to write some tests that actually test the behavior of our model.

Adding Todo model unit tests

Our Todo model is quite simple, but it's still very valuable to cover our basics by testing the following:

- The model is correctly initialized with the correct attributes (this will prevent any accidental changes to the model's structure and defaults).
- The model is correctly validated.

Let's proceed!

Update the `todos_test.jl` file and make it look like this:

```julia
using Test, SearchLight, Main.UserApp, Main.UserApp.Todos
using SearchLight.Validation

@testset "Todo unit tests" begin
  t = Todo()

  @testset "Todo is correctly initialized" begin
    @test t.todo == ""
    @test t.completed == false
  end

  @testset "Todo validates correctly" begin
    @testset "Todo is invalid" begin
      v = validate(t)
      @test haserrors(v) == true
      @test haserrorsfor(v, :todo) == true
      @test errorsfor(v, :todo)[1].error_type == :not_empty
    end
    @testset "Todo is valid" begin
      t.todo = "Buy milk"
      v = validate(t)
      @test haserrors(v) == false
      @test haserrorsfor(v, :todo) == false
      @test errorsfor(v, :todo) |> isempty == true
    end
  end
end;
```

We have defined multiple (nested) test sets that group testing logic by area of focus. Within each test set, we can define multiple tests that cover the Todo models' initialization and validation. The `"Todo is invalid"` test set passes because a freshly initialized `Todo()` has no description, and we have a validator to test that it is not empty. Verify that all 8 tests pass as we have done previously.

Interacting with the DB

Notice that we haven't touched the DB yet. Let's add a few tests to make sure that our to-do items can be correctly persisted. Let's create a new file in the test/ folder, named todos_db_test.jl, and write the following code inside it:

```
using Test, SearchLight, Main.UserApp, Main.UserApp.Todos
using SearchLight.Validation, SearchLight.Exceptions

@testset "Todo DB tests" begin
  t = Todo()

  @testset "Invalid todo is not saved" begin
    @test save(t) == false
    @test_throws(InvalidModelException{Todo}, save!(t))
  end
end;
```

If we run the tests now, we'll see that all 10 tests pass. To make our tests execute faster, we can run just the tests we're currently working on. For example, here, we run only one test file:

```
julia --project runtests.jl todos_db_test
```

Our tests confirm that invalid todos are not persisted to the DB. Now, let's make sure that the valid ones do. Append the following code above the final end part of the todos_db_test.jl file:

```
@testset "Valid todo is saved" begin
  t.todo = "Buy milk"
  @test save(t) == true
  tx = save!(t)
  @test ispersisted(tx) == true
  tx2 = findone(Todo, todo = "Buy milk")
  @test pk(tx) == pk(tx2)
end
```

We have set the todo field of our model to the "Buy milk" string, making it valid. Then, we attempt to save it to, and retrieve it from, the DB, making sure that it was correctly persisted. Let's run the tests.

Oh no, our new tests are failing! Fortunately, the error is quite easy to debug:

```
Got exception outside of a @test
  SQLite.SQLiteException("no such table: todos")
```

Of course, we previously configured our DB connection, but we haven't actually initialized the test DB, nor did we run the migrations. In addition to setting up the test DB, we must also make sure that we clean up this DB after we finish our tests; otherwise, future tests will find a DB that contains data from previous tests.

Add the following code, which sets up the test DB, right before the first `@testset "Todo DB tests"` part of our `todos_db_test.jl` file:

```
try
  SearchLight.Migrations.init()
catch
end
SearchLight.config.db_migrations_folder =
abspath(normpath(joinpath("..", "db", "migrations")))
SearchLight.Migrations.all_up!!()
```

Also, add the following code at the end, which cleans up the test DB at the end of the test run:

```
SearchLight.Migrations.all_down!!(confirm = false)
```

Verify that all tests now pass.

Adding integration tests

Next, let's take a look at how to add a few integration tests, checking that the interactions between the views, the controller, and the model are correct.

Let's add a new file to the `test` folder, named `todos_integration_test.jl`, and add the following code inside it:

```
using Test, SearchLight, Main.UserApp, Main.UserApp.Todos
using Genie
import Genie.HTTPUtils.HTTP

try
  SearchLight.Migrations.init()
catch
```

```
end

cd("..")
SearchLight.Migrations.all_up!!()
Genie.up()                                               # 1
@testset "TodoMVC integration tests" begin
  @testset "No todos by default" begin
    response = HTTP.get("http://localhost:8000/")        #2
    @test response.status == 200
    @test contains(String(response.body), "Nothing to do")
  end


  t = save!(Todo(todo = "Buy milk"))
  @testset "Todo is listed" begin
    response = HTTP.get("http://localhost:8000/")
    @test response.status == 200
    @test contains(String(response.body), "Buy milk")
  end


  @test t.completed == false
  @testset "Status toggling" begin
    HTTP.post("http://localhost:8000/todos/$(t.id)/toggle")
    @test findone(Todo, id = t.id).completed == true
  end


  @testset "Status after deleting" begin
    HTTP.post("http://localhost:8000/todos/$(t.id)/delete")
    response = HTTP.get("http://localhost:8000/")
    @test contains(String(response.body), "Nothing to do")
  end
end

Genie.down()
SearchLight.Migrations.all_down!!(confirm = false)
cd(@__DIR__)
```

As before, first, we configure the DB access. For integration tests, the application needs to be running with the web server started by calling Genie.up – and that's what we do on line *1*. Also, we need to simulate browser-server interactions using the HTTP package web client to make requests to the server, such as on line *2*. Our tests call some of the URLs defined by our app and check that the responses are correct, looking for relevant text in the response body. Run the new tests as follows:

```
julia --project runtests.jl todos_integration_test
```

They should all pass.

Finally, we can run all our tests to make sure that everything is working correctly:

```
julia --project runtests.jl

todos_db_test: .....
todos_integration_test: .......
todos_test: ........



Test Summary: | Pass  Total    Time
TodoMVC tests |   20     20   23.3s
```

Now our app has been properly tested!

Summary

In this chapter, we finally made a complete application in the Julia web framework, which you could take as a blueprint for starting your own project. You learned how to generate a new MVC project and its model, how to create views to display the data, and how to route browser requests through a controller function to its intended view. Also, you saw how you can add data validation functions and set up automatic testing for your app.

In the next chapter, we'll apply the same techniques to create a REST API-based app.

Further reading

If you need to read up on MVC design, consult the following great intro: `https://reinteractive.com/posts/310-mvc-and-rails`.

If you have a question, first consult the excellent Genie documentation at `https://genieframework.com/docs/genie/v5/`.

If your question hasn't been answered, it surely will be within a short time if you post it onto Genie's Discord channel at `https://discord.gg/9zyZbD6J7H`.

Adding a REST API

In the previous chapter, we learned how to build a Todo MVC web app using Genie. This approach uses a JS web frontend for our users. But what if we want to make our data, which is stored in a database, available to other kinds of graphical frontends, such as a mobile app or a desktop application? We don't want to rewrite everything from scratch for each new frontend. Besides, other companies may want to write these frontends to our data.

In this chapter, we'll add a REST API to our existing app by organizing our code so as to allow other applications and software to access and interact with our data. Here are the topics we are going to discuss in this chapter:

- Extending a web app with a REST API
- Writing tests for a REST API
- Documenting our API with the Swagger UI

Technical requirements

The source code for this chapter can be found at `https://github.com/PacktPublishing/Web-Development-with-Julia-and-Genie/tree/main/Chapter5`. Please make sure you use *Genie v5.11* or higher.

When writing this chapter, *SwagUI v0.9.1* and *Swagger Markdown v0.2.0* were used.

Extending a web app with a REST API

Just like the regular Julia libraries, REST APIs are meant to be used by developers in order to build applications that are interoperable. Let's first lay some groundwork.

Preparing for a REST API

Each application changes, and it is common to introduce changes in our data model or API, such as different endpoints, different HTTP methods, different request parameters, different responses and response structures, and so on. These changes could cause other apps that access our data via the REST API to crash. That's why we need versioning.

API versioning

By introducing versioning in our API, we allow consumer apps to continue working while providing their developers the time to update to the latest version.

There are a few ways to version REST APIs, and we'll use what's arguably the most popular approach: versioning through the URI path. This means that we'll place the major version of our API into the URL of the requests, as in the following link: mytodos.com/api/v1/todos. This is the preferred approach for Genie apps as it's easy to use and understand. At the same time, this approach promotes a modular architecture, as we're about to see in the next section.

Architecting our API

Taking into account the versioning requirements, our API requests will be prefixed with /api/v1, indicating that the current major version of the API is v1. In the future, if we introduce breaking changes into our API, we'll need to introduce a new major version of the API (v2, v3, and so on).

Under the hood, each part of the URI will be implemented into a distinct Julia module, making our code modular and composable for easier maintenance and extensibility. By encapsulating the API logic as well as the specific version into dedicated modules and submodules, we make our code future-proof and easy to maintain.

Defining our routes

The REST API will expose similar endpoints as the web application itself. We want to allow the consumers of our API to create, retrieve, update, and delete todos. The only difference is that we will not include a dedicated toggle endpoint because REST APIs by convention have a different update mechanism. That being said, let's define our routes by adding the following code at the end of our routes.jl file:

```
route("/api/v1/todos", TodosController.API.V1.list, method =
GET)
route("/api/v1/todos/:id::Int", TodosController.API.V1.item,
method = GET)
route("/api/v1/todos", TodosController.API.V1.create, method =
POST)
```

```
route("/api/v1/todos/:id::Int", TodosController.API.V1.update,
method = PATCH)
route("/api/v1/todos/:id::Int", TodosController.API.V1.delete,
method = DELETE)
```

The implementation plan

The code we just added defines five routes to handle the main four operations of the API: listing, creating, updating, and deleting todos. REST API best practices prescribe that we use dedicated HTTP methods for each of these operations. In addition, we'll also use JSON to handle both requests and responses. Finally, to be good citizens of the web, we'll also extend our current set of features to add support for pagination to the list of todos.

Listing todos

Let's begin with the first operation: the retrieval and listing of to-do items. Add this at the bottom of the `TodosController.jl` file, right above the closing end:

```
### API
module API
module V1

using TodoMVC.Todos
using Genie.Router
using Genie.Renderers.Json
using ....TodosController

function list()
  all(Todo) |> json
end

function item()
end

function create()
end

function update()
```

```
end

function delete()
end

end # V1
end # API
```

In the preceding snippet, we define a new module called `API` and a submodule, `V1`. Inside `V1`, we declare references to various dependencies with `using` statements. More importantly, we bring into scope `Genie.Renderers.Json`, which will do all the heavy lifting for building JSON responses for our API. You can think of it as the counterpart of `Genie.Renderers.Html`, which we used in `TodosController` to generate HTML responses. And just like in the main controller, we'll leverage the features in `Genie.Router` to handle the requests data. We have also included a reference to the main `TodosController` module, using a relative namespace two levels up (notice the four dots, `....`).

Finally, we define placeholder functions for each of the operations, matching the handlers we defined in our routes file. The `list` function even includes a bit of logic, to allow us to check that everything is set up well. Try it out at `http://localhost:8000/api/v1/todos`, and you should get a JSON response with the list of todos.

Here is our output:

```
[{"id":{"value":1},"todo":"Build the Genie TodoMVC app","completed":false},{"id":{"value":2},"todo":"Getting
groceries","completed":false},{"id":{"value":3},"todo":"Visiting my therapist","completed":true}]
```

Figure 5.1 – List of todos as JSON response

Now it's time to refine our todos listing. It's important to keep our code DRY and reuse as much logic as possible between the HTML rendering in `TodosController.jl` and our API that outputs JSON. Right now, `TodosController.index` includes both the filtering and retrieval of the to-do items, as well as the rendering of the HTML. The filtering and retrieval operations can be reused by the API, so we should decouple them from the HTML rendering.

Replace the `TodosController.index` function with the following code:

```
function count_todos()
    notdonetodos = count(Todo, completed = false)
    donetodos = count(Todo, completed = true)
```

```
  (
    notdonetodos = notdonetodos,
    donetodos = donetodos,
    alltodos = notdonetodos + donetodos
  )
end

function todos()
  todos = if params(:filter, "") == "done"
    find(Todo, completed = true)
  elseif params(:filter, "") == "notdone"
    find(Todo, completed = false)
  else
    all(Todo)
  end
end

function index()
  html(:todos, :index; todos = todos(), count_todos()...,
       ViewHelper.active)
end
```

Here, we refactor the index function to only handle the HTML rendering, while we prepare the data in two new functions: count_todos and todos. We'll reuse these functions to prepare the JSON response for our API. It's worth noticing the flexibility of Julia in the way we pass the keyword arguments to the html function inside index(): we explicitly pass the todos keyword argument with the todos() value, and we splat the NamedTuple received from count_todos() into three keyword arguments, and finally, we pass the ViewHelper.active value as the last implicit keyword argument.

Next, we can use these newly created functions in our API.V1.list function, to retrieve the actual data:

```
function list()
  TodosController.todos() |> json
end
```

We can check that our refactoring hasn't broken anything by checking some URLs for both the app and API:

- `http://localhost:8000/`

- `http://localhost:8000/?filter=notdone`

- `http://localhost:8000/api/v1/todos`

- `http://localhost:8000/api/v1/todos?filter=done`

We haven't forgotten about our integration tests and you're welcome to run those as well, but they will be more useful once we add tests for the API too.

Adding pagination

I'm hoping that our to-do list will not be that long as to actually need pagination, but pagination is a common and very useful feature of REST APIs and it's worth seeing how it can be implemented, especially as SearchLight makes it very straightforward. We want to accept two new optional query parameters, page and limit, to allow the consumer to paginate the list of todos. The page parameter will indicate the number of the page (starting with 1) and limit will indicate the number of todos per page.

As mentioned, the implementation is extremely simple; we only need to pass the extra arguments, with some reasonable defaults, to the `SearchLight.all` function that we use to get the todos. Update the `else` branch in the `TodosController.todos` function as follows:

```
function todos()
  todos = if params(:filter, "") == "done"
    find(Todo, completed = true)
  elseif params(:filter, "") == "notdone"
    find(Todo, completed = false)
  else
    # this line was: all(Todo) and is changed to:
    all(Todo; limit = params(:limit,
      SearchLight.SQLLimit_ALL) |> SQLLimit,
          offset = (parse(Int, params(:page, "1"))-1) *
                    parse(Int, params(:limit, "0")))
  end
end
```

All we need to do is pass the `limit` and `offset` arguments to the `all` function, and we're done. Given that these are optional (the users can make requests without pagination), we also set some good defaults: if there is no `limit` argument, we include all the todos by passing the `SearchLight.SQLLimit_ALL` constant to the `limit` argument. As for `offset`, this indicates how many items to skip, which are calculated by multiplying the page number by the number of items per page. If there is no page argument, we start with the first page by using 1 as the default, but do note that when we calculate the offset, we use page – 1; this way, on page 1 the offset is 0. This is because the `offset` argument in the database query represents the number of todos to skip, and for page 1, we want to skip 0 todos. As for `limit`, the default here is 0 (meaning that if no `limit` argument is passed, we'll include all the todos with `offset` of 0).

We can test the new functionality by getting a couple of pages and limiting the number of todos per page to one:

- `http://localhost:8000/api/v1/todos?page=1&limit=1`
- `http://localhost:8000/api/v1/todos?page=2&limit=1`

Also, no pagination will return all the todos, as expected (`http://localhost:8000/api/v1/todos`).

Our web app does not support pagination yet. We'll skip it, but if you want, you can do it as an exercise: add a new element to the UI to allow the users to navigate between pages of todos and set the limit to a reasonable constant value, such as 20.

Creating todos

We want to allow the consumer of our API to add new todos. We already have the route and the corresponding route handler, so it's now time to add the actual code. Add the following code snippets in this section while changing the `API.V1.create` function, as follows:

```
using Genie.Requests
using SearchLight.Validation
using SearchLight
```

In the preceding code, we first declare that we'll be using three extra modules. `Genie.Requests` provides a higher-level API to handle requests data, and we'll rely on it to help us work with the JSON payloads. The other is `SearchLight.Validation`, which we've already seen in action and which helps us to validate the data we receive from the consumer of the API. SearchLight itself gives us access to the `save!` method:

```
function check_payload(payload = Requests.jsonpayload())
    isnothing(payload) && throw(JSONException(status =
```

```
      BAD_REQUEST, message = "Invalid JSON message received"))
  payload
end
```

Given that we expect a JSON payload, in the preceding `check_payload`, we verify whether the body of the request can be converted to JSON. We use the `Requests.jsonpayload` function to do that. If the payload is not valid JSON, the `Requests.jsonpayload` function will return nothing. In this case, we throw an exception, informing the user that the message received is not valid JSON.

Now we get to the `create` function:

```
function create()
  payload = try
    check_payload()
  catch ex
    return json(ex)
  end

  todo = Todo(todo = get(payload, "todo", ""),
    completed = get(payload, "completed", false))
  persist(todo)
end
```

Here, once we are sure that we have received a valid JSON payload, we parse it, looking for relevant data to create a new to-do item. We provide some good defaults and create a new instance of our Todo model, using the provided payload. We then attempt to persist the newly created model to the database, by passing it to the `persist` function, where we apply our model validations:

```
function persist(todo)
  validator = validate(todo)
  if haserrors(validator)
    return JSONException(status = BAD_REQUEST,
      message = errors_to_string(validator)) |> json
  end

  try save!(todo)
    json(todo, status = CREATED, headers = Dict("Location"
      => "/api/v1/todos/$(todo.id)"))
  catch ex
```

```
        JSONException(status = INTERNAL_ERROR,
            message = string(ex)) |> json
    end
  end
```

In the `create` function, we first check whether the request payload is valid by invoking the `check_payload` function.

If our `persist` function finds any validation errors, we return an exception with the error details. If the validations pass, we save the todo in the database and return a JSON response with the newly created todo and the **CREATED** status code. As a best practice, we also pass an additional location header, which is the URL of the newly created todo. If for some reason the todo could not be saved, we return an exception with the error details.

We can test the various scenarios using an HTTP client such as *Postman* (for an example of how to use Postman, see *Chapter 2, Testing the ToDo Services with Postman*) or *Paw*. But we'll skip that for now and just add integration tests in the final section of this chapter.

Updating todos

Updating todos is a breeze, especially as we've already implemented our validation logic. First, we need to change the `try/catch` statement in the `persist` function to the following:

```
try
    if ispersisted(todo)
      save!(todo)
      json(todo, status = OK)
    else
      save!(todo)
      json(todo, status = CREATED, headers =
        Dict("Location" => "/api/v1/todos/$(todo.id)"))
    end
  catch ex
    JSONException(status = INTERNAL_ERROR,
        message = string(ex)) |> json
    end
```

This change allows us to detect whether the todo we're attempting to save was already persisted to the database or not. If it was, we'll update the todo in the database, otherwise, we'll create a new one, and we need to return the correct response, based on the database operation we performed.

Now, fill in our empty `API.V1.update` function as follows:

```
function update()
  payload = try
    check_payload()
  catch ex
    return json(ex)
  end
  todo = findone(Todo, id = params(:id))
  if todo === nothing
    return JSONException(status = NOT_FOUND,
      message = "Todo not found") |> json
  end

  todo.todo = get(payload, "todo", todo.todo)
  todo.completed = get(payload, "completed",
    todo.completed)

  persist(todo)
end
```

We start by checking whether the payload is valid. If it is, we continue by retrieving the corresponding todo from the database, using the `id` passed as part of the URL. If the todo is not found, we return an exception. Otherwise, we update the todo with the provided data again, applying some good defaults (in this case, keeping the existing value if a new value was not provided). Finally, we attempt to persist the todo in the database.

Deleting todos

The last operation that our API should support is the deletion of the to-do items. We'll update the `API.V1.delete` function as follows:

```
function delete()
  todo = findone(Todo, id = params(:id))
  if todo === nothing
```

```
    return JSONException(status = NOT_FOUND,
      message = "Todo not found") |> json
  end

  try
    SearchLight.delete(todo) |> json
  catch ex
    JSONException(status = INTERNAL_ERROR,
      message = string(ex)) |> json
  end
end
```

The code attempts to retrieve the todo from the database, based on the `id` passed as part of the URL. If the todo is not found, we return an exception. Otherwise, we delete the todo from the database and return it.

Retrieving todos

For retrieving individual to-do items from the database, we only need to check that the corresponding to-do item exists by looking it up by `id`. If it does not exist, we return a 404 error. If it does, we return the todo. Here is the code for the `item` function:

```
function item()
  todo = findone(Todo, id = params(:id))
  if todo === nothing
    return JSONException(status = NOT_FOUND,
      message = "Todo not found") |> json
  end

  todo |> json
end
```

Writing tests for a REST API

> **Note**
>
> The complete code listing for the tests in this section can be found at https://github.com/PacktPublishing/Web-Development-with-Julia-and-Genie/blob/main/Chapter5/TodoMVC/test/todos_API_test.jl.

It's time to see our API in action by writing a test suite to check all the endpoints and the various scenarios we've implemented. Let's start by adding a new test file for our API:

```
julia> touch("test/todos_API_test.jl")
```

Next, we'll add the test suite to our newly created file:

```julia
using Test, SearchLight, Main.UserApp, Main.UserApp.Todos
using Genie
import Genie.HTTPUtils.HTTP
import Genie.Renderers.Json.JSONParser.JSON3

try
  SearchLight.Migrations.init()
catch
end


cd("..")
SearchLight.Migrations.all_up!!()
Genie.up()

const API_URL = "http://localhost:8000/api/v1/todos"

@testset "TodoMVC REST API tests" begin

  @testset "No todos by default" begin
    response = HTTP.get(API_URL)
    @test response.status == Genie.Router.OK
    @test isempty(JSON3.read(String(response.body))) ==
      true
  end

end

Genie.down()
SearchLight.Migrations.all_down!!(confirm = false)
cd(@__DIR__)
```

Besides the declaration of the used dependencies, the first and the last parts of the file are the setup and teardown of the tests, just like we did in the integration tests in *Chapter 4, Building an MVC ToDo App*, in the *Testing Genie apps* section. This is where we set up the test database and the API server while at the end, we remove the test data and stop the web server.

All our tests will be placed inside a main testset called `"TodoMVC REST API tests"`. And our first test simply checks that when initiating our test suite, our database does not contain any todos. We make a `GET` request to our `/todos` endpoint that lists the to-do items, and we verify that the response is a `200 OK` status code and that the response body is empty.

Next, let's add tests for to-do creation. These will verify all the assumptions related to to-do creation. Append these code snippets under the `"No todos by default"` testset:

```
@testset "Todo creation" begin
  @testset "Incorrect content-type should fail todo creation"
begin
    response = HTTP.post(API_URL, ["Content-Type" =>
      "text/plain"], JSON3.write(Dict("todo" => "Buy
        milk")); status_exception = false)
    @test response.status == Genie.Router.BAD_REQUEST
    @test JSON3.read(String(response.body)) ==
      "Invalid JSON message received"
  end
```

This first test verifies that when we send a request with an incorrect content type, the response has a 400 BAD_REQUEST status code and that the response body equals the `"Invalid JSON message received"` error message:

```
  @testset "Invalid JSON should fail todo creation" begin
    response = HTTP.post(API_URL, ["Content-Type" =>
      "application/json"], "Surrender your data!";
        status_exception = false)
    @test response.status == Genie.Router.BAD_REQUEST
    @test JSON3.read(String(response.body)) ==
      "Invalid JSON message received"
  end
```

The second test checks that when we send a request with an invalid JSON payload, the API responds in the same manner, with a BAD_REQUEST status and the same error message:

```
@testset "Valid JSON with invalid data should fail todo
          creation" begin
  response = HTTP.post(API_URL, ["Content-Type" =>
    "application/json"], JSON3.write(Dict("todo" => "",
      "completed" => true)); status_exception = false)
  @test response.status == Genie.Router.BAD_REQUEST
  @test JSON3.read(String(response.body)) == "Todo should not
be empty"
end
```

The third test checks that despite the valid content type and JSON payload, if the todo data is not valid, the request will fail with a BAD_REQUEST status and the "Todo should not be empty" error message.

Moving on to the next API test:

```
@testset "No todos should've been created so far" begin
  response = HTTP.get(API_URL)
  @test response.status == Genie.Router.OK
  @test isempty(JSON3.read(String(response.body))) ==
    true
end
```

The fourth test makes an extra check that because of the previous error responses, no todos have been created up to this point:

```
@testset "Valid payload should create todo" begin
  response = HTTP.post(API_URL, ["Content-Type" =>
    "application/json"], JSON3.write(Dict("todo" => "Buy
      milk")))
  @test response.status == Genie.Router.CREATED
  @test Dict(response.headers)["Location"] ==
    "/api/v1/todos/1"
  @test JSON3.read(String(response.body))["todo"] ==
    "Buy milk"
end
```

```
@testset "One todo should be created" begin
  response = HTTP.get(API_URL)
  @test response.status == Genie.Router.OK
  todos = JSON3.read(String(response.body))
  @test isempty(todos) == false
  @test length(todos) == 1
  @test todos[1]["todo"] == "Buy milk"

  response = HTTP.get("$API_URL/1")
  @test response.status == Genie.Router.OK
  todo = JSON3.read(String(response.body))
  @test todo["todo"] == "Buy milk"
end

end # "Todo creation"
```

Finally, the last two tests confirm that when we send a valid payload, the API successfully creates a new todo, returns it with a 201 Created status code, and the location header is set to the new todo's URL and that we can retrieve it.

Next, for the todo updating tests, use the following code (see the link for the complete code at the start of this section):

```
@testset "Todo updating" begin

  @testset "Incorrect content-type should fail todo
    update" begin
    response = HTTP.patch("$API_URL/1", ["Content-Type"
      => "text/plain"], JSON3.write(Dict("todo" => "Buy
        soy milk")); status_exception = false)
...

  end
```

The following tests are exactly the same as the tests for creating todos, except that the request is `HTTP.patch("$API_URL/1",...)` instead of `HTTP.post(API_URL,...)`, because PATCH is needed for an update:

```
@testset "Invalid JSON should fail todo update"


@testset "Valid JSON with invalid data should fail todo
   update"
```

The code for the following testset is the same as for the `"One todo should be created"` testset:

```
@testset "One existing todo should be unchanged"
```

The following testsets have almost the same code as their counterparts from the create todos testset:

```
@testset "Valid payload should update todo"
@testset "One existing todo should be changed"
```

The following test is new:

```
@testset "Updating a non existing todo should fail"
  begin
  response = HTTP.patch("$API_URL/100", ["Content-Type"
    => "application/json"], JSON3.write(Dict("todo" =>
       "Buy apples")); status_exception = false)
  @test response.status == Genie.Router.NOT_FOUND
  @test JSON3.read(String(response.body)) ==
    "Todo not found"
end


end # "Todo updating"
```

These tests follow the logic of the todo creation testset, just adapted to the todo updating scenario, so we won't get into details about these.

Now, let's add the todo deletion tests (for the complete code, see the repository linked at the beginning of this section).

The most significant change is that the request is now of the `HTTP.delete("$API_URL/1")` form:

```
@testset "Todo deletion" begin
```

```
@testset "Deleting a non existing todo should fail" begin
  response = HTTP.delete("$API_URL/100", ["Content-Type"
    => "application/json"]; status_exception = false)
  @test response.status == Genie.Router.NOT_FOUND
  @test JSON3.read(String(response.body)) ==
    "Todo not found"
end
```

The logic should be pretty clear by now. The first test checks that when we try to delete a non-existing todo, the API responds with a NOT_FOUND status and the "Todo not found" error message:

```
@testset "One existing todo should be deleted" begin
  response = HTTP.delete("$API_URL/1")
  @test response.status == Genie.Router.OK
  @test JSON3.read(String(response.body))["todo"] ==
    "Buy vegan milk"
  @test HTTP.get("$API_URL/1"; status_exception =
    false).status == Genie.Router.NOT_FOUND
end
```

The second test checks that when we delete an existing todo, the API responds with an OK status and the todo data.

At this point there should be no todos left in the database:

```
@testset "No todos should've been left" begin
  response = HTTP.get(API_URL)
  @test response.status == Genie.Router.OK
  @test isempty(JSON3.read(String(response.body))) ==
    true
end
end # "Todo deletion"
```

The last test in this testset makes sure that no todos are left in the database.

And finally, to complete our test suite, we'll add the pagination tests:

```
@testset "Todos pagination" begin
  todo_list = [
    Dict("todo" => "Buy milk", "completed" => false),
```

```
      Dict("todo" => "Buy apples", "completed" => false),
      Dict("todo" => "Buy vegan milk", "completed" => true),
      Dict("todo" => "Buy vegan apples", "completed" =>
          true),
  ]


  for todo in todo_list
    response = HTTP.post(API_URL, ["Content-Type" =>
      "application/json"], JSON3.write(todo))
  end
```

The preceding code is a bit more involved. First, we create a to-do list, which comprises some fake data to mock our tests. Next, we iterate over this list and use the API itself to create all the todos. Once our data is in, it's time for the actual tests:

```
  @testset "No pagination should return all todos" begin
    response = HTTP.get(API_URL)
    @test response.status == Genie.Router.OK
    todos = JSON3.read(String(response.body))
    @test isempty(todos) == false
    @test length(todos) == length(todo_list)
  end
```

The first test checks that when we don't specify any pagination parameters, the API returns all todos.

Now, we test the outputted data with various pagination scenarios, making sure that the data is split correctly between the various pages, according to the limit parameter:

```
  @testset "One per page" begin
    index = 1
    for page in 1:length(todo_list)
      response = HTTP.get("$API_URL?page=$(page)&limit=1")
      todos = JSON3.read(String(response.body))
      @test length(todos) == 1
      @test todos[1]["todo"] == todo_list[index]["todo"]
      index += 1
    end
  end
```

For the last four testsets, see the code in the repository:

```
@testset "Two per page"
@testset "Three per page"
@testset "Four per page"
@testset "Five per page"
end # "Todos pagination"
```

Now, to verify these tests, go to your command-line interface, and within the `test` folder, issue the following command:

```
julia --project runtests.jl todos_API_test
```

You'll see something like the following on your display:

```
todos_API_test: ....................................................
............................


Test Summary: |  Pass  Total    Time
TodoMVC tests |   78     78   23.6s
```

Now verify the complete testset as we have done in *Chapter 4*, in the *Testing Genie apps* section. You should have 98 tests that pass.

Documenting our API with the Swagger UI

Having built our REST API and tested it, it is very important to document the API, because you can have a potentially unknown number of customers who will want to use it. Swagger (`https://swagger.io/`) is a very useful tool to help you design and document your APIs. Swagger UI employs the `OpenAPI` standard and allows us to document our API in code, and at the same time, to publish it via a web-based human-readable interface.

In order to add support for Swagger UI, we need to add two new packages to our project, `SwagUI` and `SwaggerMarkdown`:

```
pkg> add SwagUI, SwaggerMarkdown
```

We will set up the Swagger comments and the API documentation functionality in the `routes.jl` file. The routes for the web application remain the same, but the API routes are now augmented with `swagger"..."` annotations that are used to build the API documentation. Add `using SwagUI, SwaggerMarkdown` to the `routes.jl` file, and update the REST API routes as follows:

> **Note**
>
> Because it is quite lengthy, we are only showing the complete docs code for the `GET` request of the `/apI/v1/todos` route in this section. You can find the complete code at `https://github.com/PacktPublishing/Web-Development-with-Julia-and-Genie/blob/main/Chapter5/TodoMVC/routes.jl`.

```
swagger"
/api/v1/todos:
  get:
    summary: Get todos
    description: Get the list of todos items with their
                 status
    parameters:
      - in: query
        name: filter
        description: Todo completed filter with the values
                     'done' or 'notdone'
        schema:
          type: string
          example: 'done'
      - in: query
        name: page
        description: Page number used for paginating todo
                     items
        schema:
          type: integer
          example: 2
      - in: query
        name: limit
        description: Number of todo items to return per
                     page
        schema:
```

```
            type: integer
            example: 10
    responses:
      '200':
        description: A list of todos items
  post:
    "
route("/api/v1/todos", TodosController.API.V1.list, method =
GET)
route("/api/v1/todos", TodosController.API.V1.create, method =
POST)


swagger"
/api/v1/todos/{id}:
  get:
      patch:
      delete:
    "
route("/api/v1/todos/:id::Int", TodosController.API.V1.item,
method = GET)
route("/api/v1/todos/:id::Int", TodosController.API.V1.update,
method = PATCH)
route("/api/v1/todos/:id::Int", TodosController.API.V1.delete,
method = DELETE)


### Swagger UI route

route("/api/v1/docs") do
  render_swagger(
    build(
      OpenAPI("3.0", Dict("title" => "TodoMVC API",
              "version" => "1.0.0")),
    ),
    options = Options(
      custom_favicon = "/favicon.ico",
      custom_site_title = "TodoMVC app with Genie",
      show_explorer = false
```

```
        )
    )
end
```

In the preceding code, we have first grouped the routes by path, differentiating them by method. We have two distinct paths, `/api/v1/todos` and `/api/v1/todos/:id`. The first path accepts `GET` and `POST` requests to list and create todos, while the second path accepts `GET`, `PATCH`, and `DELETE` requests to retrieve, update, and delete a to-do item.

The Swagger documentation is built by annotating the individual paths and sub-differentiating them by method. Then, for each path and method combination, we detail the request and response information, including properties such as `summary`, `description`, `requestBody`, and `responses`.

In addition, at the end of the file, we now have a new route to render Swagger UI. This route invokes the `render_swagger` function, passing various configuration options to build the docs.

This was all! Restart your app by following the steps from the *Pausing development* subsection in *Chapter 4*. Our API is now documented and we can use Swagger UI to browse the API by accessing the `/api/v1/docs` route at `http://localhost:8000/api/v1/docs`. Not only that but the browser is fully interactive, allowing us to run queries against the API and see the results in real time.

Here is the main documentation screen:

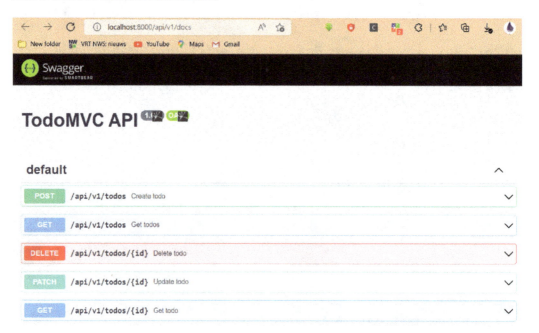

Figure 5.2 – Swagger documentation screen

Summary

In this chapter, we added a REST API to our Todo application. You learned how to code the CRUD functions for this API and about the necessity of versioning your code. The REST API exposes your data to the world, that's why testing and documenting are especially important. You learned how to write these specific tests and how to document the API using Swagger.

In the next chapter, we'll discover the specific techniques to deploy our Genie app in production.

Further reading

Another related and interesting topic is GraphQL, a new query language and runtime for APIs (see `https://graphql.org/`). There is a Genie example for GraphQL here: `https://github.com/neomatrixcode/Diana.jl/tree/master/samples/genie`.

6

Deploying Genie Apps in Production

A key part of the web development workflow is being able to deploy an app in production environments and making sure it runs in a secure and performant manner. Genie, together with some of the packages available in the Genie ecosystem, provides a multitude of useful features for deploying and running applications in demanding production environments, with a focus on performance, stability, and security.

You will learn how to use separate development, test, and production environments, how to optimize your app for production, and how to set up your app for production implementation using **Docker** and **Heroku**. At the end of the chapter, you'll know everything you need to deploy your app in a production environment.

In this chapter, we will discuss the following topics:

- Genie app environments
- Containerizing Genie apps with Docker
- Setting up our GitHub repository for **continuous integration (CI)**
- Deploying Genie apps with Git and Docker containers
- Improving application startup time with `PackageCompiler.jl`
- Deploying our optimized app on Heroku
- Wrapping up the benefits of Docker

Technical requirements

To follow along, the following software needs to be installed on your computer:

- **Git**: This app allows you to work with GitHub repositories. Install it from `https://git-scm.com/downloads`.

- **Docker**: This application enables you to package your app into a container that contains everything it needs to run, including the OS. Download and install the appropriate Docker Desktop for your system from `https://docs.docker.com/get-docker/`.

The source code for this chapter can be found at `https://github.com/PacktPublishing/Web-Development-with-Julia-and-Genie/tree/main/Chapter6`.

Genie app environments

Genie applications run in the context of an *environment*, which is a way of configuring a Genie application with a group of settings that are optimized for a certain task, such as development, testing, or high-performance execution in production. In other words, we can define multiple environments, each with its specific configuration, and then we can easily swap the environment to enable all the corresponding settings at once.

Out-of-the-box Genie apps come with three environments: `dev` (which stands for development), `prod` (for production), and `test` (for testing). Each environment has its own configuration file with the same name, placed inside the `config/env/` folder of the app. These environments come with preconfigured settings for running tasks optimized for three common situations: development (with rich logging and code reload and re-compilation defaults), testing, and high-performance production runtime.

The environment that is used the most by developers is `dev`, the default environment that the app uses that is optimized for running the application for development. `dev` provides features that make the development process more efficient and productive, such as the following:

- Code reloading and recompilation every time files are saved by automatically setting up file loading with the `Revise.jl` package

- Extensive and rich error messages and error stacks

- Automatic serving of assets such as images, stylesheets, and scripts

- The `dev` environment also has sensible settings for running the application locally, such as using the `127.0.0.1` host and the default Genie port, `8000`

However, development features such as code reloading, extensive logging and detailed error messages, or asset serving are not appropriate when we run the application in production, either because these features slow down the application or because they can expose sensitive information that can be exploited by attackers.

For such situations, we use the `prod` environment, which provides configurations that are optimized for running the application in production. The `prod` environment disables code reloading and recompilation, disables detailed error messages, and recommends the disabling of asset serving. In addition, production apps will by default use the host `0.0.0.0`, which is usually what's expected when deploying on most hosting platforms.

Finally, the third bundled environment, `test`, is optimized for testing the application, and we've already seen it in action in *Chapter 4, Building an MVC ToDo App*, under the *Testing Genie apps* section.

Customizing the environments

We can edit the environment files in order to change, remove, or add configuration elements. Take, for instance, the default `dev.jl` file:

```
using Genie, Logging
Genie.Configuration.config!(
    server_port                  = 8000,
    server_host                  = "127.0.0.1",
    log_level                    = Logging.Info,
    log_to_file                  = false,
    server_handle_static_files   = true,
    path_build                   = "build",
    format_julia_builds          = true,
    format_html_output           = true,
    watch                        = true
)


ENV["JULIA_REVISE"] = "auto"
```

The `config!` method modifies and returns the `Genie.config` object, which is an instance of `Genie.Configuration.Settings` and represents the application's configuration. You can probably recognize here some of the configurations we have already mentioned, such as, for instance, the host and the port of the application, the logging settings, handling of assets (static files), or various formatting options that are useful in development.

We can also use the environment files to add environment-dependent settings, such as, for instance, the JULIA_REVISE configuration, which sets automatic file recompilation when files change by employing the Revise.jl package.

By contrast, take a look at the default prod.jl file:

```
using Genie, Logging
Genie.Configuration.config!(
  server_port                = 8000,
  server_host                = "0.0.0.0",
  log_level                  = Logging.Error,
  log_to_file                = false,
  server_handle_static_files = true, # (1)
              path_ build = "build",
  format_julia_builds        = false,
  format_html_output         = false
)


if Genie.config.server_handle_static_files
  @warn("For performance reasons Genie should not serve static
files (.css, .js, .jpg, .png, etc) in production.
      It is recommended to set up Apache or Nginx as a
reverse proxy and cache to serve static assets.")
end


ENV["JULIA_REVISE"] = "off"
```

In the preceding code, we can see the differences in the server configuration (host and port), logging, formatters, and automatic recompilation. As indicated in line *(1)*, for performance reasons, Genie should not serve static files (.css, .js, .jpg, .png, etc.) in production.

It is recommended to set up **Apache** or **Nginx** as a reverse web proxy and cache to serve static assets.

Creating extra environments

The three default environments cover some of the most common use cases, but we can define other environments as needed. For instance, many development teams commonly use a staging environment as an intermediary stage between development and production. All we need to do in order to enable a new environment is to create the corresponding env file. For instance, we can create a copy of our `prod.jl` file and name it `staging.jl` to define a staging environment, modifying it as necessary:

```
# config/env/staging.jl
using Genie, Logging

Genie.Configuration.config!(
    server_port                  = 8000,
    server_host                  = "0.0.0.0",
    log_level                    = Logging.Debug,
    log_to_file                  = true,
    server_handle_static_files   = true, # for best
performance set up Nginx or Apache web proxies and set this to
false
    path_build                   = "build",
    format_julia_builds          = true,
    format_html_output           = true
)

ENV["JULIA_REVISE"] = "off"
```

The preceding snippet shows a possible staging configuration where we keep some of the production settings but enable more comprehensive logging and some extra formatting to help us debug potential issues before we release the application in production.

SearchLight database environments

Equally important is the ability to automatically configure the database connection based on environments. SearchLight integrates with Genie's environments to automatically pick the right database connection. This is very important in order to avoid accidentally polluting or destroying production data when we run our application in the development or testing environment.

Remember that we have already configured a distinct `test` database in our `db/connection.yml` file in *Chapter 4, Configuring the Test Database*:

```
env: ENV["GENIE_ENV"]

dev:
  adapter:  SQLite
  database: db/dev.sqlite3

test:
  adapter:  SQLite
  database: db/test.sqlite3
```

Note   how, at the top of the preceding code file, we set `env` to automatically pick the application's environment, which in turn allows SearchLight to connect to the corresponding database.

Changing the active environment

In the section about unit tests (see *Adding Todo model unit tests* in *Chapter 4*), we have seen how the very first thing in the `test/runtests.jl` file (our test runner) is to change the environment of the application to test: `ENV["GENIE_ENV"] = "test"`.

Now we understand why this is important: to apply the right configuration during tests and to connect to the right database.

As such, one way of changing the application's environment is by passing the environment's name as a Julia environment variable, either by setting it in the ENV global or by passing it as a command-line argument when starting the app. We'll see in just a minute how to switch our application to run in production, but before we can do that, there's one more thing we need to do: prepare the database.

We have not yet defined a database configuration for our `prod` environment, and this will cause the app to error out at startup. So, let's make sure we add it first. Append the following to the end of the `db/connection.yml` file:

```
prod:
  adapter:  SQLite
  database: db/prod.sqlite3
```

SearchLight will create the `prod.sqlite3` database the next time we start the app in the `prod` environment.

Starting the application in production

By default, Genie apps start in development, as that is the logical first step once an app is created: to develop it. But we can easily change the active environment at any time. However, this must be done when the app is started in order to allow the proper loading of the environment's settings. Changing the environment when the app is running always requires restarting the app.

Using environment variables

One way to change the active environment is by passing the app's active env as a command-line environment variable. Environment variables are key-value pairs, stored by Julia in the ENV collection, that offer information about the current context of the Julia execution. We can access these variables from within our app as ENV["<variable_name>"]. We can define our environment variables when starting our app, by passing them as extra command-line arguments.

For instance, we can configure our Genie app to not show the Genie loading banner and overwrite the web server port by starting our app with the following command in a terminal (be sure to start it from within the app's root folder).

On Windows, run:

```
$ bin/server GENIE_BANNER=false PORT=9999
```

On Linux/macOS, run:

```
$ GENIE_BANNER=false PORT=9999 bin/server
```

This command will disable the Genie banner and will start the application on port 9999, producing the following output:

```
Ready!
┌ Info: 2022-08-07 16:21:56
└ Web Server starting at http://127.0.0.1:9999 - press Ctrl/
Cmd+C to stop the server.
```

In the same way, we can pass the GENIE_ENV environment variable to our script in order to start the app with the designated environment; for example, for production, we would run the following.

On Windows:

```
$ bin/server GENIE_ENV=prod
```

On Linux/macOS:

```
$ GENIE_ENV=prod bin/server
```

For testing, we would run the following.

On Windows:

```
$ bin/repl GENIE_ENV=test
```

On Linux/macOS:

```
$ GENIE_ENV=test bin/repl
```

Using config/env/global.jl

You may have noticed that in the `config/env` folder, there is a `global.jl` file that by default only contains a comment. As the comment indicates, we can use this file to define and apply global configuration variables – that is, settings that will be applied to all environments. Think of it as a way to avoid copying the same settings in all the environment files.

However, as this file is loaded right before the specific environment file for the app, we can actually use it to change the active environment. For instance, if we add this line to the `global.jl` file, our application will always run in a `prod` environment:

```
ENV["GENIE_ENV"] = "prod"
```

Setting the active environment in the `global.jl` file will always overwrite the configuration set via GENIE_ENV on the command line.

Running the app in production

Let's restart our app now in production, for example, by using the GENIE_ENV environment variable:

```
$ bin/repl GENIE_ENV=prod
```

Upon restarting the app in production, our database was automatically created, but SearchLight has only created an empty DB. We need to set up the database structure by running database migrations:

```
julia> using SearchLight
julia> SearchLight.Migration.init()
julia> SearchLight.Migration.allup()
```

Now, everything is ready for our app to run in production. We can test it by starting the server with `julia> up()` and visiting `http://localhost:8000`. Our to-do app should run as expected – but of course, you won't be able to see any of the to-do items you may have added in development as, in production, the app is using the new production DB. You'll find the to-do items when restarting the app in dev mode again. This level of data isolation provided by application environments ensures that we don't accidentally run development or test code using the production data.

With our app fully configured to run in production, we're now ready to deploy it on the internet for production use. That's what we will do in the next sections.

Containerizing Genie apps with Docker

Docker deployments are the most common way of releasing and scaling web applications as part of DevOps workflows. Genie has official support for Docker containerization via the `GenieDeployDocker` plugin. Let's use it to containerize our app.

We'll start by adding the `GenieDeployDocker` package:

```
pkg> add GenieDeployDocker
```

Once installed, we'll use it to generate a Dockerfile for our application, which is the configuration file that tells Docker how to containerize our app:

```
julia> using GenieDeployDocker

julia> GenieDeployDocker.dockerfile()
Docker file successfully written at /path/to/your/app/TodoMVC/
Dockerfile
```

If you're familiar with Docker, you can take a look at the resulting Dockerfile. Right out of the box, it contains everything that is needed to set up a Linux container with preinstalled Julia, set up our application and its dependencies, and start the server to listen on the designated ports. You can read more about the Dockerfile in the official Docker documentation at `https://docs.docker.com/engine/reference/builder/`.

We'll need to make only one change in the Dockerfile; toward the bottom of the file, there is a line that reads `ENV GENIE_ENV "dev"`. This sets the environment used by the app. By default, it's set to `dev`; edit this line and set the app's environment to `prod`.

Now that we have a Dockerfile, we can ask Docker to build our container:

```
julia> GenieDeployDocker.build()
```

This process can take a bit of time as Docker will pull the Linux OS image from the internet, install and precompile our app's dependencies, copy our application into the Linux container, and finally, run the app by starting the server. As you run the `build` command, you'll be able to follow the progress of the various steps as the REPL's output.

Once the build finishes, we can *deploy* our application in the Docker container locally, that is, run the container and access the application within the container running on our computer. Let's do it to confirm that everything works as expected:

```
julia> GenieDeployDocker.run()
```

The preceding code will start our Genie application inside the Docker container, in the production environment, by running the `bin/server` script – as configured by the line `CMD ["bin/server"]` in the Dockerfile. In addition, it will bind the app's port inside the container (port 8000) to port 80 of the Docker host, which is your computer. This means that, after the familiar Genie loading screen, on confirming that the application is ready, you can access it by simply visiting `http://localhost` in your browser.

Until now, we have run our tests manually. This could easily be forgotten at some point. Let's make GitHub do this work for us automatically!

Setting up our GitHub repository for CI

In this step, we'll set up a GitHub repository for our *TodoMVC* app. We'll use GitHub for two main actions: to set up continuous integration (CI) and have GitHub Actions run our test suite every time we push to the repository, and to serve as a public repository that we can access from our deployment servers.

For the following actions, you will need a free GitHub account. Log in to your GitHub account and create a new repository to host the app at `https://github.com/new`. Give it a good name, such as `GenieTodoMVC`. Put in a description too if you want, then click on **Create repository**.

Once the GitHub repository is created, we need to configure your local Genie app to use it. Going back to your computer, in the terminal, in the app's folder, run the following (you will need to have `git` installed on your computer):

```
$ git init
$ git add .
$ git commit -m "initial commit"
$ git branch -M main
$ git remote add origin <HTTPS URL OF YOUR GITHUB REPO>
$ git push -u origin main
```

Once our app's code has been pushed to GitHub, we can set up our CI workflow to take advantage of our test suite. This integration will automatically run every time we push code to our repository.

Inside the root of our app, create a new folder named `.github`, and inside this, create a new folder named `workflows`. Next, within the `workflows` folder, create a new file, `ci.yml`, and add the following content to it:

```yaml
name: ci
on:
  - push
  - pull_request
jobs:
  test:
    name: Julia ${{ matrix.version }} - ${{ matrix.os }} - ${{ matrix.arch }}
    runs-on: ${{ matrix.os }}
    strategy:
      fail-fast: false
      matrix:
        version:
          - '1.6'
          - '1.7'
          - '1.8'
          - 'nightly'
        os:
          - ubuntu-latest
          - macOS-latest
          - windows-latest
        arch:
          - x64
    steps:
      - uses: actions/checkout@v2
      - uses: julia-actions/setup-julia@latest
        with:
          version: ${{ matrix.version }}
          arch: ${{ matrix.arch }}
      - uses: julia-actions/julia-buildpkg@latest
      - uses: julia-actions/julia-runtest@latest
```

This configuration file will run our test suite on every `git push` and `git pull` request, on four Julia versions (*1.6*, *1.7*, *1.8* and *nightly*), on the three main operating systems. Our testing strategy covers all the relevant Julia versions: *1.6* which is LTS, *1.7* which is the previous minor version that is still widely used, *1.8* which is the current stable version, and *1.9* which is nightly (upcoming version that is currently being developed and will be released in the future).

When you finish don't forget to push the changes to GitHub:

```
$ git add .
$ git commit -m "CI"
$ git push -u origin main
```

That's it, now our application is fully configured for CI on GitHub.

Deploying Genie apps with Git and Docker containers

Now that we have confirmed that our application runs correctly in a Docker container, we can deploy our application on any of the multitude of web hosting services that support Docker container deployments. By using Docker containers, we can be sure that the exact setup described in the Dockerfile and tested on our machine will be run and configured on the hosting service.

AWS EC2 hosting

AWS is the most popular hosting platform at the moment so let's see how to deploy our Genie app there. AWS has a multitude of services (over 100) providing a huge array of possible deployment setups. Most of the AWS configurations are quite complex and go beyond the scope of this chapter, with large books and month-long certification programs being dedicated to teaching AWS usage. We'll go with one of the simplest and most straightforward ways to get the application up and running.

To follow along with the next section, you will need a free AWS account – a credit card is required in order to open the AWS account. Follow these steps:

1.   Start by going to `https://signin.aws.amazon.com` and logging in – if you don't have an account already, sign up.

2.   Once you sign in to the AWS console, go to the EC2 dashboard (`https://console.aws.amazon.com/ec2/v2/home`) and click on **Launch instances**.

3. In the **Launch Instance** wizard, first, give the instance a name, such as `GenieTodoMVC`.

Figure 6.1 – Naming the launch instance

4. Then, for the OS image, search for `Amazon linux 2 ami`. From the search results, pick the 64-bit (x86) **Amazon Linux 2 AMI (HVM) - Kernel 5.10, SSD Volume Type** (or newer kernel version if available at the time of reading). We'll use the **t2.micro** with 1 vCPU and 1 GB of RAM. This image is free tier eligible, meaning that if you qualify for the free tier offer, you'll use this for free.

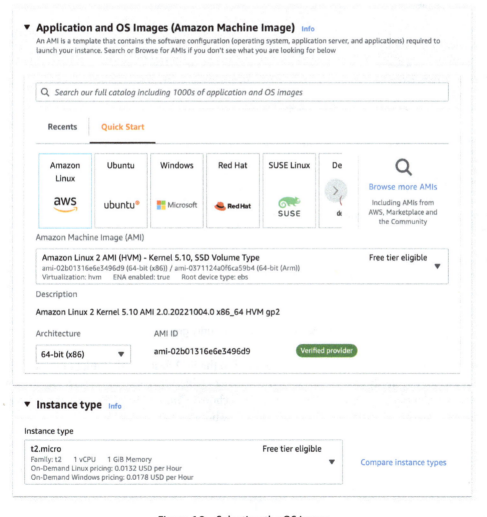

Figure 6.2 – Selecting the OS image

5. Next, create an SSH key pair; to keep things simple, we won't use it now, but download it and store it safely so you can log in to your server over SSH in the future.

6. Then, go ahead and create a security group – leave the SSH access and make sure to allow HTTP and HTTPs traffic from the internet.

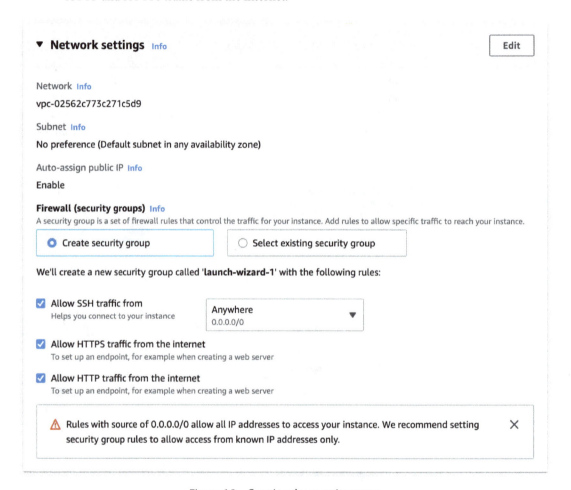

Figure 6.3 – Creating the security group

That's it, we can now launch the instance and go back to the EC2 dashboard. It might take a couple of minutes to see our newly created server in the instances table. Once it's visible, select its row and click **Actions | Connect** (or right-click on the row and choose **Connect** from the contextual menu). This will open a terminal into the EC2 instance in a new browser tab.

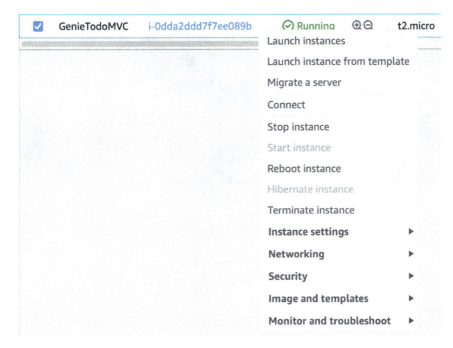

Figure 6.4 – EC2 instance in a new browser tab

Installing and configuring Docker

Now that our web server is online, we need to install Docker and Git to run our deployment workflow:

```
$ sudo yum install -y docker
$ sudo yum install -y git
```

With Docker and Git installed, let's make sure that our Docker service runs correctly, as by default it is disabled:

```
$ sudo service docker status
```

If this command reports the status as Active: inactive (dead), it means that the Docker service has been installed but it hasn't been correctly started. We'll start it manually to make sure that all goes well:

```
$ sudo service docker start
```

If there are no errors, we can check the status again – we should see `Active: active (running)`.

```
[ec2-user@ip-172-31-45-94 ~]$ sudo service docker status
Redirecting to /bin/systemctl status docker.service
● docker.service - Docker Application Container Engine
   Loaded: loaded (/usr/lib/systemd/system/docker.service; disabled; vendor preset: disabled)
   Active: inactive (dead)
     Docs: https://docs.docker.com
[ec2-user@ip-172-31-45-94 ~]$ sudo service docker start
Redirecting to /bin/systemctl start docker.service
[ec2-user@ip-172-31-45-94 ~]$ sudo service docker status
Redirecting to /bin/systemctl status docker.service
● docker.service - Docker Application Container Engine
   Loaded: loaded (/usr/lib/systemd/system/docker.service; disabled; vendor preset: disabled)
   Active: active (running) since Wed 2022-11-09 17:57:29 UTC; 4s ago
```

Figure 6.5 – Successful active status

Now, set Docker to autostart:

```
$ sudo systemctl enable docker
$ sudo usermod -aG docker ec2-user
```

Finally, restart the EC2 instance:

```
$ sudo reboot
```

This can take a couple of minutes; after restarting, just reconnect back to the server from the EC2 dashboard as described previously in the *AWS EC2 hosting* section.

Cloning the GitHub repository

With Docker up and running, it's time to clone our app's source code onto the server. We'll use our public GitHub repository. Run this into the EC2 instance terminal:

```
$ git clone <URL TO YOUR GITHUB REPO>
```

If you're having problems with accessing your repository, you can use the public repository we created while writing the book, available at `https://github.com/essenciary/GenieTodoMVC`.

Next, move into the app's directory that we just cloned: `cd <YOUR REPO NAME>` (for example, `$ cd GenieTodoMVC`).

Now it is time to build our Docker container:

```
$ docker build . -t todomvc
```

Once the build completes, we can run our container, mapping our app's port (8000) to port 80 (HTTP) of our EC2 instance:

```
$ docker run -d -p 80:8000 todomvc
```

Now the application will be accessible on the public IPv4 address as well as on the public IPv4 DNS indicated for your instance in the EC2 dashboard.

> **Note**
>
> You need to access the application over HTTP, not HTTPS, as we have not configured an SSL certificate for our app. Setting up SSL certificates on AWS for EC2 instances goes beyond the scope of this chapter but you can find the information by reading the various guides and tutorials that are publicly available (see, for example, `https://docs.aws.amazon.com/AWSEC2/latest/UserGuide/SSL-on-amazon-linux-2.html`).

Setting up the production database on AWS

Our application runs well, however, there is an issue with our current configuration. Because we're using an SQLite database, our database is now inside the container, which exposes our data to being lost if our container is destroyed. In addition, SQLite is not the best choice for production databases, for instance, due to limitations when it comes to concurrent writes. Finally, in general, the best practice is to set up the database outside the application's container so that they won't compete over resources, especially in a high-load scenario.

As we're using AWS, let's employ one of the available cloud database services. We'll use **Relational Database Service** (**RDS**), which gives us access to managed relational databases in the AWS cloud, including commonly used backends such as MySQL/MariaDB, PostgreSQL, and Oracle. SearchLight supports all three of these, so we have multiple choices – we'll go with **MariaDB**. Go through the following steps:

1. Start by visiting the RDS home page at `https://eu-west-3.console.aws.amazon.com/rds/home` and clicking on **Create database**.

2. In the next step, pick **Standard create** for the database creation method, **MariaDB** for the engine options, and **Free tier** for the template.

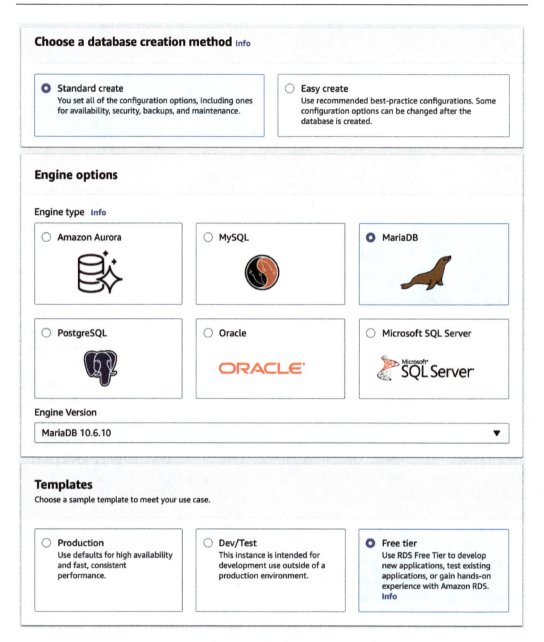

Figure 6.6 – Choosing a database creation method

3. Then, in the **Settings** section, you can give a name to the instance (database server) using the **DB Instance Identifier** field, and set up the master username and the master password. Make sure to write down the username and password as we'll need them to connect.

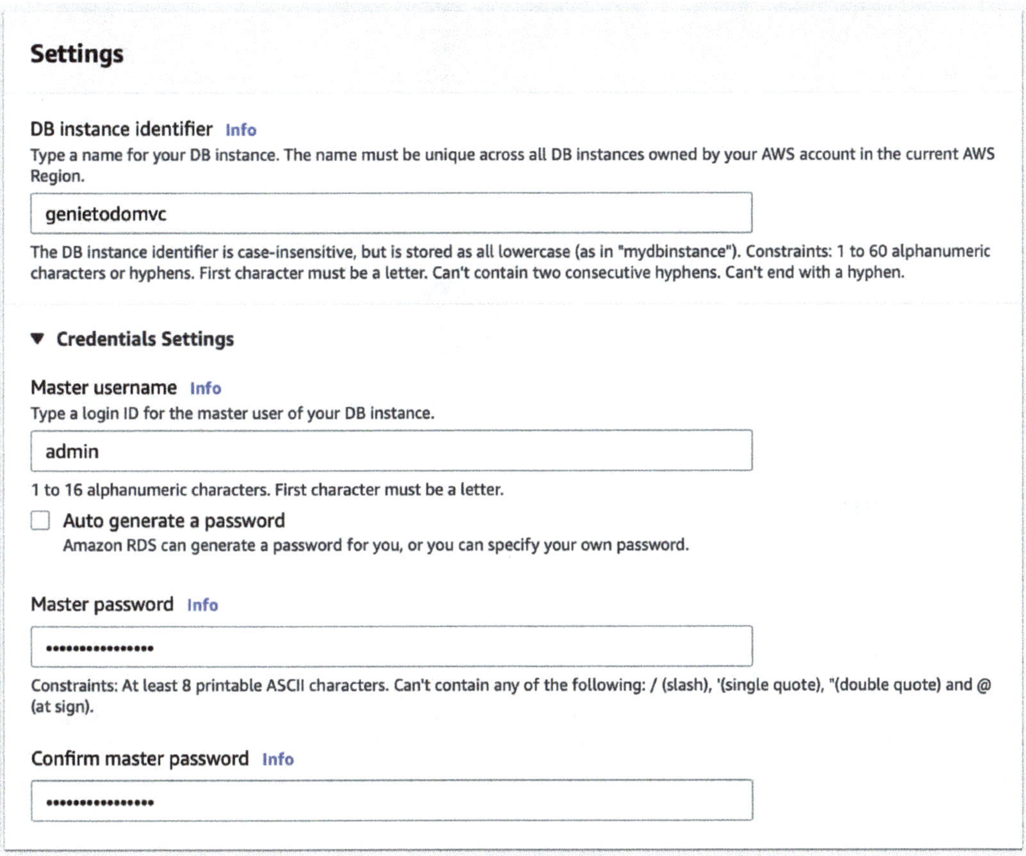

Settings

DB instance identifier Info
Type a name for your DB instance. The name must be unique across all DB instances owned by your AWS account in the current AWS Region.

```
genietodomvc
```

The DB instance identifier is case-insensitive, but is stored as all lowercase (as in "mydbinstance"). Constraints: 1 to 60 alphanumeric characters or hyphens. First character must be a letter. Can't contain two consecutive hyphens. Can't end with a hyphen.

▼ **Credentials Settings**

Master username Info
Type a login ID for the master user of your DB instance.

```
admin
```

1 to 16 alphanumeric characters. First character must be a letter.

☐ Auto generate a password
     Amazon RDS can generate a password for you, or you can specify your own password.

Master password Info

```
••••••••••••••••
```

Constraints: At least 8 printable ASCII characters. Can't contain any of the following: / (slash), '(single quote), "(double quote) and @ (at sign).

Confirm master password Info

```
••••••••••••••••
```

Figure 6.7 – Setting up the username and password

4. Leave the rest of the options as default until you get to **Public access** in the Connectivity section - and set public access to **Yes**.

5. Then, for **VPC security group**, leave **Choose existing** and pick the security group you have already set up for the web app.

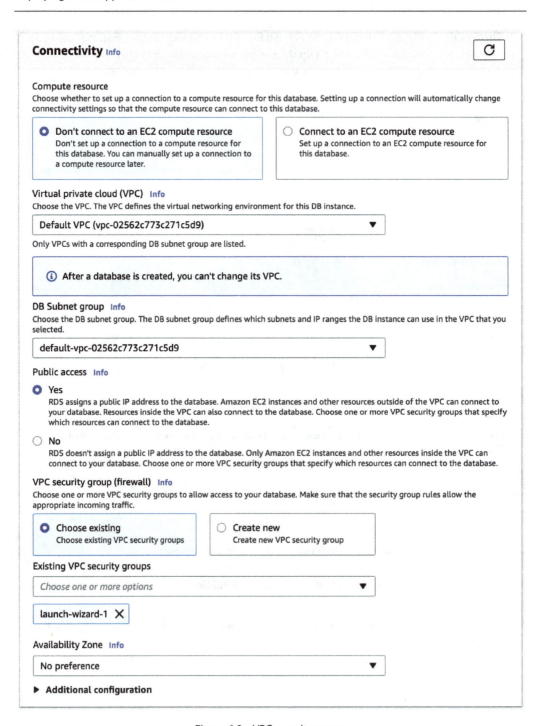

Figure 6.8 – VPC security group

6. For **Database authentication**, leave in **Password authentication**, then expand **Additional configuration** and put a name for **Initial database name**.

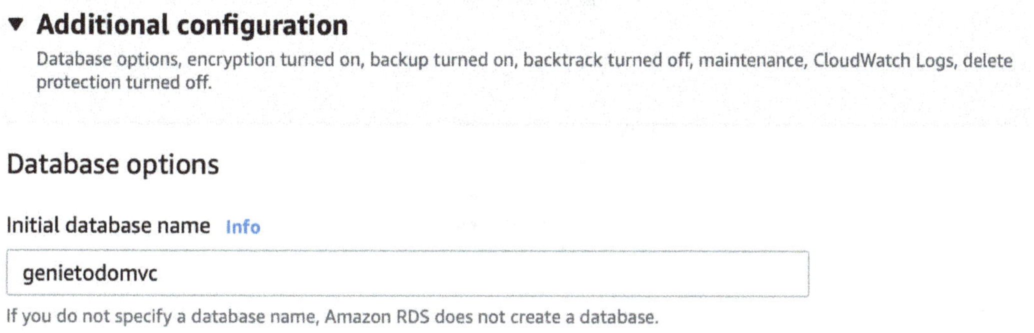

▼ **Additional configuration**
   Database options, encryption turned on, backup turned on, backtrack turned off, maintenance, CloudWatch Logs, delete protection turned off.

Database options

Initial database name   Info

genietodomvc

If you do not specify a database name, Amazon RDS does not create a database.

Figure 6.9 – Adding the initial database name

7. Leave the rest of the options as their defaults and click **Create database** at the bottom of the page.

After this, you will be redirected to the dashboard page: `https://console.aws.amazon. com/rds/home?#databases`. You may have to wait a few minutes for the database to be ready, as indicated in the **Status** column showing **Creating**.

Once the database instance becomes available, you can click on it to see its details. In the **Connectivity and security** section, you will find the endpoint and the port that can be used to connect to the DB. We can now test that our DB is set up and accessible by connecting from our computer, using a MySQL client. If you don't have one, you can try `DbGate`, a free database client supported on all major operating systems (`https://dbgate.org/database/mysql-client.html`).

Open the MySQL client and configure it to use the **endpoint** as the host. Use the default port `3306`, and input the master user and master password for username and password. If you have the option to configure the default database, put the name of the DB you have configured when setting up the RDS instance. If all went well, you will be able to confirm the correct setup by successfully connecting to the RDS database.

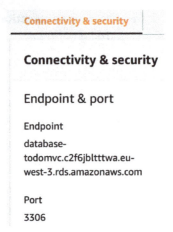

Figure 6.10 – Confirmation of correct setup

Preparing our production app to use the RDS database

Now that we have configured our database, we need to ensure that it can be used by our app. There are two things we need to address: connecting to the RDS database, and ensuring that all database migrations are run.

Automating database migrations

Let's start with the migrations. We want to make sure that all the migrations are automatically run in production for every build we release. The simplest way to do this is to configure migrations to be run every time the application is started, and we can achieve this by adding the following code at the bottom of the `config/initializers/searchlight.jl` file:

```
try
    SearchLight.Migration.init()
catch
end
SearchLight.Migration.allup()
```

Now, every time the app starts, it will ensure that the migrations are configured and that all the available migrations are up. The `Migration.init` method will throw an exception if it has already been run, so we put it in a `try/catch` block. The `Migration.allup` function, however, will not throw any exception if the migrations are already up; it will simply not run any migrations if all are up.

Configuring the RDS database connection

Now it's time to configure our application to use the RDS database. You might be tempted to just go and add the connection info to the db/connection.yml file. This can be a viable option in most situations, but given our flow, where we use a public GitHub repository, it's a bad idea. Our db/ connection.yml is pushed to the GitHub repository, exposing the connection info to our publicly accessible database, meaning that anybody would be able to connect to our DB! Instead, we'll pass the database connection info as environment variables for our Docker container on AWS.

What we need to configure though is the fact that we'll use the MySQL adapter, and we can also safely set the name of the database:

1. So, comment out or delete the current prod connection that uses SQLite and add the following MySQL connection:

    ```
    prod:
      adapter:  MySQL
      database: <name of your database>
    ```

2. We also need to add SearchLightMySQL as a dependency of our app, otherwise, SearchLight won't be able to connect to the MariaDB backend. In the app's Julia REPL, run the following:

    ```
    (GenieTodoMVC) pkg> add SearchLightMySQL
    ```

3. When finished, push the changes to GitHub:

    ```
    $ git commit -am "MySQL support and autorun migrations"
    $ git push -u origin main
    ```

4. Next, go to **AWS | EC2 Dashboard | Instances** at https://console.aws.amazon. com/ec2/v2/home?#Instances:instanceState=running, and connect to the EC2 instance we created earlier (the web server).

5. Once connected, check whether the container is running and if it is, get its name:

    ```
    $ docker container ls
    ```

6. Stop the running container using the following:

    ```
    $ docker stop <name of container>
    ```

7. Now, make sure that you move into the application's folder, for example, cd GenieTodoMVC, and pull the changes from the GitHub repository with git pull.

8. Almost done. The last step is to make the database information available to our container. There are two ways to do this.

    •   One way to do this is to pass all the info as environment variables to the Docker command, like this:

```
SEARCHLIGHT_USERNAME=<master username> SEARCHLIGHT_
PASSWORD=<master password> SEARCHLIGHT_HOST=<database
endpoint> docker run -d -p 80:8000 todomvc
```

    However, this is a bit verbose and creates a potential security issue as the command, including the database login info, would be stored in the terminal's history.

    •   We're better off using another Docker feature, namely the env-file option. This allows us to pass an environment text file that includes all the connection data. Let's create this env file and use it. We'll use the nano text editor, which should already be available on the Linux instance (if it's not, add it with sudo yum install nano):

```
$ nano ../env.list
```

9. The nano editor will create the file and open it up for editing. Type in the following content, putting in your actual connection data:

```
SEARCHLIGHT_HOST=<database endpoint>
SEARCHLIGHT_USERNAME=<master username>
SEARCHLIGHT_PASSWORD=<master password>
```

10. Save the file (*Ctrl + O*) and exit (*Ctrl + X*).

11. That's all – now we can start the server by running the Docker container:

```
$ docker run -d -p 80:8000 --env-file=../env.list todomvc
```

12. It might take a bit of time to start. If the Docker operations take too long, it is recommended to increase the size of the EC2 instance by changing its type, for example to t2.large. This will incur a small cost so remember to stop the instance after you're done with testing the configuration.

    If you want to peek into the running logs of the app, you can check the container's logs as follows:

```
$ docker container ls
```

13. Get the name of the container and then run the following:

```
$ docker logs <name of container>
```

Once the app is ready, you can access it on the container's public IPv4 address or on the public IPv4 DNS, as listed on the EC2 instance summary page.

> **Important note**
> Remember to use HTTP not HTTPS, as we have not configured an SSL certificate for our app. Setting up SSL certificates on AWS for EC2 instances goes beyond the scope of this chapter but you can find the information by reading the various guides and tutorials that are publicly available.

Autogenerating the secrets.jl file

If you looked at the Docker container's logs, you may have noticed that the production app shows a warning that **No secret token is defined**. The secret token is a unique random sequence of characters that is different for each Genie app and is used to encrypt data used by the application, such as sessions and cookies. This token is stored in the `config/secrets.jl` file, which is by default added to `.gitignore`, meaning that it won't be pushed to our GitHub repository and it won't be pulled onto our server. The technique helps us avoid accidentally pushing sensitive data to public GitHub repositories.

The problem, however, is that if the `secrets.jl` file is missing, Genie will generate a temporary one and use it to encrypt the data. Every time the app is restarted, a new secret token is generated, and when it changes, data encrypted with a different secret can't be decrypted. So, let's extend our application to make sure that production apps automatically generate their secret file, by adding the following line at the bottom of `config/env/prod.jl`:

```
Genie.Secrets.secret_file_exists() || Genie.Generator.write_
secrets_file()
```

Because Julia uses **JIT** compilation, the application's code is automatically compiled while the application is running, the compilation being triggered as needed, every time a piece of code that has not been already compiled is invoked. As such, when an application is started, a large part of the codebase will need to be compiled. Understandably, this initial compilation time, during which the application is unresponsive, can be a problem – and even more so for a web application, where response times are critical. To remedy this potential problem, the `PackageCompiler.jl` library can help us out.

Improving application startup time using a custom sysimage

It's important to understand that we're talking only about the initial response times after the application is started when most of the codebase is JIT compiled. This is known in Julia parlance as *time to first plot*. Once the initial compilation is completed, the application will run and respond very fast, which is a great feature for web applications that can run for weeks and months between restarts.

Thanks to the efforts of the Julia stewards and the community, the *time to first plot* has been going down, and work is being done to allow ahead-of-time compilation for Julia apps. Meanwhile, one of the best solutions available today is to use `PackageCompiler.jl` (https://github.com/JuliaLang/PackageCompiler.jl) to create a custom Julia library, called a **sysimage**, that is optimized for our specific application, to reduce the startup latency of our app.

Technically, this process is about creating a custom sysimage. The details of this process are beyond the scope of this chapter but you can read about it at https://julialang.github.io/PackageCompiler.jl/stable/sysimages.html.

We'll use our Dockerfile to define the steps for generating the sysimage so that it's automatically created each time we build our app in the Docker container:

1. Edit the Dockerfile and add the following lines of code after `RUN useradd --create-home --shell /bin/bash genie`:

```
# C compiler for PackageCompiler
RUN apt-get update && apt-get install -y g++
```

This simply instructs Docker to install the g++ compiler, which is needed by `PackageCompiler.jl`.

2. Then, after `RUN julia -e "using Pkg; Pkg.activate(\".\"); Pkg.instantiate(); Pkg.precompile(); "`, add the following:

```
# Compile sysimage
RUN julia --project compiled/make.jl
```

This line runs a Julia script that handles the sysimage creation process. Let's set it up.

3. Create the `compiled/` folder inside the app's directory and inside it, add the `make.jl` file with the following content:

```
using PackageCompiler

include("packages.jl")

PackageCompiler.create_sysimage(
    PACKAGES,
    sysimage_path = "compiled/sysimg.so",
    precompile_execution_file = "compiled/precompile.jl",
    cpu_target = PackageCompiler.default_app_cpu_target()
)
```

This file calls the `PackageCompiler.create_sysimage` function, passing the packages that need to be added to the sysimage, the path to where the sysimage should be saved, and the path to the precompilation file. The precompilation file is a file that runs the app to trigger the JIT compilation of our code and store the compiled parts in the sysimage.

4. Now, create the `compiled/packages.jl` file with the following content:

```
const PACKAGES = [
  "Genie",
  "HTTP",
  "Inflector",
  "Logging",
  "SearchLight",
  "SearchLightMySQL",
  "SwagUI",
  "SwaggerMarkdown"
]
```

Here, we define constant `PACKAGES` that list the packages we want to be included in the custom sysimage. These are the specific packages used by the app when running in production.

5. Add `compiled/precompile.jl` like this:

```
ENV["GENIE_ENV"] = "dev"

using Genie
Genie.loadapp(pwd())

import HTTP

@info "Hitting routes"
for r in Genie.Router.routes()
  try
    r.action()
  catch
  end
end

const PORT = 50515

try
  @info "Starting server"
  up(PORT)
catch
end
```

```
try
  @info "Making requests"
  HTTP.request("GET", "http://localhost:$PORT")
catch
end

try
  @info "Stopping server"
  Genie.Server.down!()
catch
end
```

This is a simple script that invokes the route handlers in the Hitting routes section before starting the server, making a request to the home page, and then stopping the server.

6. Next, as our application will use the custom sysimage, we must configure it to load. For this, we'll edit the bin/server file to add the --sysimage option, pointing to the location of our custom sysimage file. Make the file look like this:

```
julia --color=yes --depwarn=no --project=@.
--sysimage=compiled/sysimg.so -q -i -- $(dirname $0)/../
bootstrap.jl -s=true "$@"
```

7. Finally, going back to our app's Julia REPL, we'll need to add PackageCompiler.jl as a dependency:

```
(GenieTodoMVC) pkg> add PackageCompiler
```

If you want, you can now run build in Docker to make sure that everything is right (but beware that the sysimage generation step can take quite a long time, depending on the performance of your computer).

In the last part of the chapter, let's see how to deploy our containerized application to another hosting provider: **Heroku**.

Deploying our optimized app on Heroku

Heroku has fewer features compared to AWS and it can be more expensive to host applications there. However, it is much easier to set up and configure, providing a friendly UI for common tasks, such as setting up an SSL certificate or configuring a custom domain name.

If you don't have an account already, follow these steps:

1. Start by creating a free Heroku account by visiting `https://www.heroku.com`.

2. Genie greatly simplifies Heroku deployments thanks to the `GenieDeployHeroku.jl` package, so let's add it:

   ```
   (GenieTodoMVC) pkg> add GenieDeployHeroku
   ```

   The package uses the Heroku CLI, which needs to be installed manually. Follow the instructions for your OS from `https://devcenter.heroku.com/articles/heroku-cli#install-the-heroku-cli`.

3. Once the CLI is installed, in the app's Julia REPL, run the following:

   ```
   julia> GenieDeployHeroku.login()
   ```

4. Follow the instructions to log in to your Heroku account. Once logged in, we can create a new app instance on Heroku to host our application, with the following:

   ```
   julia> GenieDeployHeroku.createapp("<name of the app>")
   ```

5. Now we're ready to build our container:

   ```
   julia> GenieDeployHeroku.push("<name of the app>")
   ```

   This will trigger the Docker `build` process, which will also include sysimage creation.

When finished (this can take a long time – even half an hour or more, depending on the performance of your computer), we can release the application in production.

But before we do that, remember that we need to configure the database connection. Heroku offers less control over the server environment, so we can't open an SSH session and run Docker manually. However, similar features are offered by Heroku using its web UI.

To start your configuration, follow these steps:

1. Go to `https://www.heroku.com` and check your app's list: `https://dashboard.heroku.com/apps`.

2. You should see the newly created app there. Click on it to see its details.

3. Go to the **Settings** tab and click on **Reveal Config Vars**. This is the place where we can add environment variables that will be passed to the Docker process. Go ahead and add three config vars: `SEARCHLIGHT_HOST`, `SEARCHLIGHT_USERNAME`, and `SEARCHLIGHT_PASSWORD`. For the values, set the RDS database endpoint for the host, and the master username and password for the other two.

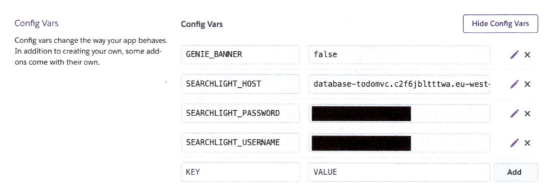

Figure 6.11 - Setting up the environmental variables

As you can see, we will be using the same database as before, but if you want to try it for yourself, Heroku also provides a similar managed cloud database service.

4. Now, we can deploy the app in production:

    ```
    julia> GenieDeployHeroku.release("<name of the app>")
    ```

5. And finally, we can open the browser to navigate to our live app using the following:

    ```
    julia> GenieDeployHeroku.open("<name of the app>")
    ```

Wrapping up the benefits of Docker

Containerized deployments using Docker are some of the most commonly used application deployment strategies today. Virtually all the modern hosting platforms provide support for Docker deployment, for basic to very complex configurations that use container orchestration frameworks such as Docker Compose or Kubernetes.

Docker deployments are very useful because by using the Dockerfile, we can implement a complex build and release workflow that will run the same everywhere. This is especially useful for Genie applications where we want to take advantage of environments and apply optimization techniques, including the building of a custom sysimage.

Custom sysimages help by greatly reducing compilation and thus compilation time, decreasing the so-called *time to first plot*, and also reducing memory and CPU needs for the app, allowing us to deploy on small and inexpensive servers. Finally, the Genie package ecosystem greatly simplifies the deployment of Genie web apps through easy-to-use deployment plugins such as `GenieDeployDocker` and `GenieDeployHeroku`.

Summary

In this chapter, we discussed the different possibilities for deploying Genie apps. First, we saw how to customize a production environment. We learned how we can package our app in a Docker container, how to set up a repository on GitHub with automatic execution of tests (CI), and how to deploy our app in a Docker container using Git. We also learned how to improve the app's startup by using `PackageCompiler.jl`. Finally, we explored an alternative way of deployment by using Heroku.

In the next chapter, we'll add authentication to our app to keep the data safe by using the GenieAuthentication plugin.

7

Adding Authentication to Our App

We've made great progress so far, developing our to-do app and hosting it on the web. However, making our application available on the internet introduces a new problem: how do we keep our data safe? Anybody who knows the URL of our app can access it and can see all our to-dos. We need to add some kind of authentication to our app so that only authorized users can access it. In addition, wouldn't it be nice if we could share our to-do app with our friends and family so they could also create their lists and keep track of their to-dos?

To achieve these outcomes, we need to add an authentication layer to our app. Authentication will ensure that only authorized users can see specific to-do items. In other words, before allowing users to create or edit to-do items, we will ask the users to authenticate themselves. If they are new to the website, they will be asked to register. If they are already registered, they will be able to use their credentials (username and password) to log in. In addition, we will also make sure that each to-do item is associated with a specific user. This way, only the user who created the to-do item will be able to see it.

Here are the topics we are going to discuss in this chapter:

- Adding and configuring the `GenieAuthentication` plugin
- Restricting access to the data
- Updating our tests

Technical requirements

The source code for this chapter can be found at `https://github.com/PacktPublishing/Web-Development-with-Julia-and-Genie/tree/main/Chapter7`.

To get all the correct versions of all the packages used in this app, go to the project root folder and open up a Julia REPL. Then enter the following commands:

```
julia> ]  # enter the package manager
pkg> instantiate
pkg> update
```

Then, do the same in the test/ folder.

Adding and configuring the GenieAuthentication plugin

The easiest way to add authentication to a Genie app is to use the **GenieAuthentication** plugin. Let's add it and follow the installation instructions (https://github.com/GenieFramework/GenieAuthentication.jl) to set up our app for authentication.

In a terminal, start the Genie REPL for the TodoMVC app. Go to the application folder and run bin/repl if you are on Linux or macOS, or bin\repl.bat if you are on Windows. Then, in the REPL, type] to enter the Pkg mode and add the GenieAuthentication plugin:

```
pkg> add GenieAuthentication
```

Once the plugin is installed, we need to configure it:

```
julia> using GenieAuthentication
julia> GenieAuthentication.install(@__DIR__)
```

By running the install function, the plugin has added all the necessary integrations into our app (views, controller, model, migrations, and so on). You can see all the operations performed by the install function by looking at the output in the REPL.

Configuring the GenieAuthentication plugin

Now that the plugin is installed, let's configure it to our needs. First, we said that we want to allow users to register, so let's enable this functionality. Registration is disabled by default as a security precaution, to make sure that we don't accidentally allow unwanted registrations on our app. To enable user registration, we need to edit the newly created genie_authentication.jl file in the plugins/ folder (this was one of the files created by the install function). Open the file and uncomment the two routes at the bottom of the file:

```
# UNCOMMENT TO ENABLE REGISTRATION ROUTES
```

```
route("/register", AuthenticationController.show_register,
named = :show_register)
route("/register", AuthenticationController.register, method =
POST, named = :register)
```

> **What are the plugins?**
>
> In case you are wondering about the plugins/ folder, it's worth mentioning that this is a *special* Genie folder. The files placed inside this folder behave very similarly to the initializer files hosted in the config/initializers/ folder. The plugins/ folder is designed to be used by Genie plugins to add their integration and initialization logic, and the only difference compared to regular initializers is that the files in the plugins/ folder are loaded after the initializers so they can get access to all the features of the Genie app (for example, the database connection, logging, and so on).

Setting up the database

The GenieAuthentication plugin stores the user information in the application's database. For this reason, we'll need to create a new table to store the user information. The plugin has created a migration file for us in the migrations/ folder. Let's run the migration to create the users table. Go back to the Genie app REPL and run the following:

```
julia> using SearchLight
julia> SearchLight.Migration.status()
```

This shows us the status of the migrations:

```
julia> SearchLight.Migration.status()
[ Info: 2022-10-05 19:50:13 SELECT version FROM schema_migrations ORDER BY version DESC
 |   | Module name & status                          |
 |   | File name                                     |
 |---|-----------------------------------------------|
 |   |               CreateTableUsers: DOWN      | |
 | 1 | 2019052410085235_create_table_users.jl       |
 |   |               CreateTableTodos: UP        |
 | 2 | 2022052910095674_create_table_todos.jl        |
```

Figure 7.1 – Showing the migration status

We can see that we have one migration, `create_table_users`, that has not been run yet. Let's run it:

```julia
julia> SearchLight.Migration.allup()
```

The `Migration.allup` function will run the migrations that have not been run yet. Alternatively, we can run a specific migration by passing its name to the `Migration.up` function; for example, in our case, it is the following: `SearchLight.Migration.up("CreateTableUsers")`.

Running the migration will create a new table in the database called `users`. The table only includes a minimum set of columns that are required by the `GenieAuthentication` plugin: `id`, `username`, `password`, `name`, and `email`. If you want to customize this structure, you can edit the migration before running it or create additional migrations.

Restricting access to the app

It's time to give our authentication feature a try. Let's go ahead and restrict access to the list of to-do items. To do this, edit the `app/resources/todos/TodosController.jl` file as follows:

1. At the top of the file, under the last `using` statement, add the following:

    ```julia
    using GenieAuthentication
    using TodoMVC.AuthenticationController
    using TodoMVC
    ```

2. Change the `index` function by adding the `authenticated!()` function call. This effectively restricts access to the body of the function to only authenticated users. The updated index function should look like this:

    ```julia
    function index()
      authenticated!()

      html(:todos, :index; todos = todos(),
            count_todos()..., ViewHelper.active)
    end
    ```

That's all we need to do for now in terms of code. However, before testing our app we need to reload it to give Genie the opportunity to load the plugin. Exit the Genie REPL and start it again (follow the instructions from the *Pausing Development* section in *Chapter 4*), then start the server with `julia> up()` and open the application in the browser (`http://localhost:8000`).

You should see the following login screen now:

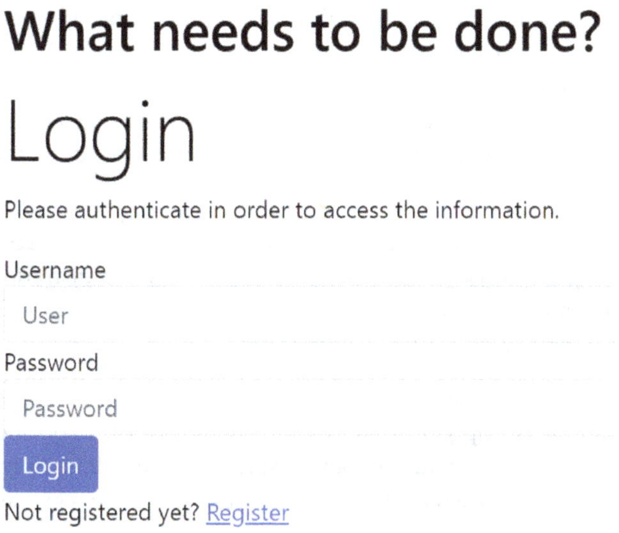

What needs to be done?

Login

Please authenticate in order to access the information.

Username

User

Password

Password

Login

Not registered yet? Register

Figure 7.2 – The login screen

Registering a new user

This time, as you can see in the preceding screenshot, we will not be able to see the list of to-dos. Instead, we will be redirected to the login page because we are not authenticated. Let's enable the registration functionality and create a new user. We enabled the registration routes earlier by uncommenting the routes. We'll need to do the same for the registration link on the login page. Open the `app/resources/authentication/views/login.jl` file and uncomment the section at the bottom of the file by deleting the first and last lines (the ones that say `Uncomment to enable registration`):

```
<!-- Uncomment to enable registration
<div class="bs-callout bs-callout-primary">
  <p>
    Not registered yet? <a
      href="$(linkto(:register))">Register</a>
  </p>
</div>
Uncomment to enable registration -->
```

After you delete the two lines and reload the page, at the bottom, under the login form, you should see a link to the registration. Clicking on the **Register** link will take us to the registration page, displaying a form that allows us to create a new account. Let's fill it up with some data and create a new user. Upon successful registration, we will get a message saying **Registration successful**. Let's improve on this by redirecting the user to their to-do list instead. Edit the `app/resources/authentication/AuthenticationController.jl` file and change the `register` function. Look for the line that says `Redirect successful` and replace it with `redirect("/?success=Registration successful")`.

Let's try out the new flow by navigating back to the registration page `http://localhost:8000/register` and creating a new user. This time, after the successful registration, the user will be automatically logged in and will be taken to the to-do list page, with the app displaying a success message, notifying them that the registration was successful.

If you want, you can also try an invalid registration – for example, by reusing the same username or by leaving some of the fields empty. You will see that the plugin will automatically guard against such issues, blocking the invalid registration and displaying a default error message indicating the problematic field. As a useful exercise, you can further improve the registration experience by customizing the error message.

Note: as we haven't added a logout button yet, you can log out by navigating to `http://localhost:8000/logout`.

Restricting access to the data

Our app is now protected by authentication, but we still need to make sure that the user can only see their own to-do items. To do this, we need to modify our app so that for each to-do item, we also store the user ID that created the to-do, effectively associating each to-do item with a user. Once we have that, we'll need to further modify our code to only retrieve the to-do items that belong to the currently logged-in user.

Adding the user ID to the to-do items

In order to associate each to-do item with a user, we need to add a new column to the `todos` table. This means we'll need to create a new migration. Let's do that by running the following command in the Genie REPL:

```
julia> using SearchLight
julia> SearchLight.Migration.new("add column user_id to todos")
```

This will create a new migration, AddColumnUserIdToTodos – let's edit it to put in our logic. In the db/migrations/ folder, open the file that ends in add_column_user_id_to_todos.jl and make it look like this:

```
module AddColumnUserIdToTodos

import SearchLight.Migrations: add_columns, remove_columns,
add_index, remove_index

function up()
  add_columns(:todos, [
    :user_id => :int
  ])

  add_index(:todos, :user_id)
end

function down()
  remove_index(:todos, :user_id)

  remove_columns(:todos, [
    :user_id
  ])
end

end
```

The migration syntax should be familiar to you by now. We are adding a new column called user_id to the todos table and a new index on that column (this is a good practice to improve the performance of queries given that we will filter the to-dos by the data in this column). The down function will undo the changes made by the up function, by first removing the index and then dropping the column. Let's run our migration:

```
julia> SearchLight.Migration.up()
```

Modifying the Todo model

Now that we have the new column in the database, we need to modify the `Todo` model to include it. Open the `app/resources/todos/Todos.jl` file and change the model definition to look like this:

```
@kwdef mutable struct Todo <: AbstractModel
    id::DbId = DbId()
    todo::String = ""
    completed::Bool = false
    user_id::DbId = DbId()
end
```

We have added a new field called `user_id` of the `DbId` type, which will be used to reference the ID of the user that created the to-do.

> **Important**
>
> Julia requires a restart when definitions of structs are changed. Exit the Genie REPL and start it again, otherwise, the application will not work correctly from this point on.

Now that we have added the column to store the user ID of the owner of the to-do item, let's update our existing to-do items to set their `user_id` to the ID of our user. This is the ID of the user that we just created during the registration process. If you want to check what users are in the database, run the following in the Genie app REPL:

```
julia> using TodoMVC.Users
julia> all(User)
```

You will get a list of all users in the database. In my case, it looks like this:

```
2-element Vector{User}:
 User
 | KEY                | VALUE                                                      |
 |--------------------|------------------------------------------------------------|
 | email::String      | adrian@geniecloud.io                                       |
 | id::DbId           | 1                                                          |
 | name::String       | Adrian                                                     |
 | password::String   | d74ff0ee8da3b9806b18c877dbf29bbde50b5bd8e4dad7a3a725000feb82e8f1 |
 | username::String   | adrian                                                     |

 User
 | KEY                | VALUE                                                      |
 |--------------------|------------------------------------------------------------|
 | email::String      | j@j.com                                                    |
 | id::DbId           | 2                                                          |
 | name::String       | John                                                       |
 | password::String   | 03ac674216f3e15c761ee1a5e255f067953623c8b388b4459e13f978d7c846f4 |
 | username::String   | john                                                       |
```

Figure 7.3 – List of users

> **Note**
>
> If you haven't created a user for yourself yet, do that now by navigating to http://localhost:8000/register and registering.

First, let's see a list of all to-dos in the database at this moment:

```
julia> using TodoMVC.Todos
julia> all(Todo)
```

Let's check the ID of our user – that is, the user that will be associated with the to-do items we previously created. In my case, the ID is 1. Now, let's update the existing to-do items to set their user_id to 1 (or whatever ID has the user you want to use). Run the following in the Genie app REPL:

```
julia > using SearchLight
julia> for t in all(Todo)
          t.user_id = 1
          save!(t)
       end
```

Now, all our existing to-dos are associated with the ID of the user. Two more things left: first, filter the to-dos by the user ID of the authenticated user when retrieving them, and second, make sure that the user ID is set when creating a new to-do item.

Getting information about the authenticated user

So far so good. Adding a user_id manually was not hard. But how do we get the user_id of the authenticated user? As it turns out, this information is readily available through the same GenieAuthentication plugin. If you check the plugins/genie_authentication.jl file, you will see that it exports only two names: current_user() and current_user_id(). The first one returns the User instance corresponding to the currently authenticated user, and the second one returns just the ID of that user (as an Int). If a user is not authenticated, both functions return nothing. We'll use the current_user_id() function to filter the to-dos by the user ID of the authenticated user. And as these functions are exported by the plugin, they are included directly and exposed by our application's main module, TodoMVC. So, we need to make sure that we add using TodoMVC to the top of app/resources/todos/TodosController.jl, as well as inside the API.V1 submodule.

Filtering the to-dos by user ID

Let's proceed by updating our application logic to filter the to-dos by the user ID of the authenticated user. Open the app/resources/todos/TodosController.jl file and make the following changes:

1. In the count_todos function, add a new filter, user_id = current_user_id(), to the count function to only count the to-dos that belong to the authenticated user:

```
function count_todos()
  notdonetodos = count(Todo, completed = false,
                         user_id = current_user_id())
  donetodos = count(Todo, completed = true,
                      user_id = current_user_id())

  (
    notdonetodos = notdonetodos,
    donetodos = donetodos,
    alltodos = notdonetodos + donetodos
  )
end
```

2. In the `todos` function, add the same filter to all the find calls:

```
function todos()
  todos = if params(:filter, "") == "done"
    find(Todo, completed = true,
         user_id = current_user_id())
  elseif params(:filter, "") == "notdone"
    find(Todo, completed = false,
         user_id = current_user_id())
  else
    find(Todo; limit = params(:limit,
         SearchLight.SQLLimit_ALL) |> SQLLimit,
              offset = (parse(Int, params(:page
                 , "1"))-1) * parse(Int,
                 params(:limit, "0")),
              user_id = current_user_id())
  end
end
```

3. Then, apply the same logic to the `toggle`, `update`, and `delete` functions:

```
function toggle()
  todo = findone(Todo, id = params(:id),
                 user_id = current_user_id())
  if todo === nothing
    return Router.error(NOT_FOUND, "Todo item with id
      $(params(:id))", MIME"text/html")
  end

  todo.completed = ! todo.completed

  save(todo) && json(todo)
end

function update()
  todo = findone(Todo, id = params(:id),
                 user_id = current_user_id())
```

```
  if todo === nothing
    return Router.error(NOT_FOUND, "Todo item with id
      $(params(:id))", MIME"text/html")
  end


  todo.todo = replace(jsonpayload("todo"), "<br>"=>"")


  save(todo) && json(todo)
end


function delete()
  todo = findone(Todo, id = params(:id),
                 user_id = current_user_id())
  if todo === nothing
    return Router.error(NOT_FOUND, "Todo item with id
      $(params(:id))", MIME"text/html")
  end

  SearchLight.delete(todo)

  json(Dict(:id => (:value => params(:id))))
end
```

4. Next, update our API module by modifying the relevant functions in the `TodosController.jl` file within the `API.V1` module, by adding the same `user_id` filter, as follows:

```
function item()
  todo = findone(Todo, id = params(:id),
                 user_id = current_user_id())
  if todo === nothing
    return JSONException(status = NOT_FOUND,
      message = "Todo not found") |> json
  end
    todo |> json
end


function update()
```

```
  payload = try
    check_payload()
  catch ex
    return json(ex)
  end

  todo = findone(Todo, id = params(:id),
                 user_id = current_user_id())
  if todo === nothing
    return JSONException(status = NOT_FOUND,
      message = "Todo not found") |> json
  end

  todo.todo = get(payload, "todo", todo.todo)
  todo.completed = get(payload, "completed",
                       todo.completed)

  persist(todo)
end

function delete()
  todo = findone(Todo, id = params(:id),
                 user_id = current_user_id())
  if todo === nothing
    return JSONException(status = NOT_FOUND,
      message = "Todo not found") |> json
  end

  try
    SearchLight.delete(todo) |> json
  catch ex
    JSONException(status = INTERNAL_ERROR,
                  message = string(ex)) |> json
  end
end
```

5. We also need to bring the authentication features into the scope of the `API.V1` submodule by adding the following `using` statements at the top of the `API.V1` module:

```
using GenieAuthentication
using TodoMVC.AuthenticationController
using TodoMVC
```

Setting the user ID when creating a new to-do item

Now that we can retrieve the to-dos by the user ID of the authenticated user, we need to make sure that the user ID is set when creating a new to-do item. In the same file, `app/resources/todos/TodosController.jl`, update the `create` function to set the `user_id`:

```
function create()
  todo = Todo(todo = params(:todo),
              user_id = current_user_id())

  validator = validate(todo)
  if haserrors(validator)
    return
      redirect("/?error=$(errors_to_string(validator))")
  end

  if save(todo)
    redirect("/?success=Todo created")
  else
    redirect("/?error=Could not save
              todo&todo=$(params(:todo))")
  end
end
```

Then, in the `API.V1` module, in the same controller file, update the `create` function to set the `user_id`:

```
function create()
  payload = try
    check_payload()
  catch ex
    return json(ex)
```

```
    end

    todo = Todo(todo = get(payload, "todo", ""),
            completed = get(payload, "completed", false),
            user_id = current_user_id())

    persist(todo)
end
```

Enhancing the validation rules

One last thing: remember that the current_user_id() function returns nothing if the user is not authenticated. This is a valid value for the user_id field, allowing us to create to-do items without being authenticated. This is not what we want, so we need to add a validation rule to the Todo model to make sure that the user_id is not nothing when creating a new to-do item.

We need to create a new validation rule in the app/resources/todos/TodosValidator.jl file. We will call it dbid_is_not_nothing. Add the following code inside the module, at the bottom, right under the is_unique function body:

```
function dbid_is_not_nothing(field::Symbol,
m::T)::ValidationResult where {T<:AbstractModel}
    isa(getfield(m, field), SearchLight.DbId) &&
        isa(getfield(m, field).value, Nothing) && return
        ValidationResult(invalid, :DbId_is_not_nothing,
        "should not be nothing")

    ValidationResult(valid)
end
```

The rule will retrieve the indicated field from the model and determine whether it's of the SearchLight.DbId type, and make sure that its value property is not nothing. If it is, it will return an invalid validation result, producing an exception – otherwise, it will return a valid result.

To enable the validation rule, in the app/resources/todos/Todos.jl file, update the model validator logic by replacing it with the following code:

```
SearchLight.Validation.validator(::Type{Todo}) =
ModelValidator([
```

```
    ValidationRule(:todo, TodosValidator.not_empty)
    ValidationRule(:user_id,
                    TodosValidator.dbid_is_not_nothing)
])
```

We have added a new validation rule, dbid_is_not_nothing, which will ensure that the user_id cannot be left unset upon saving a to-do item.

Securing all the public-facing pages

Now that we've made sure that authentication works as expected by creating a new user and logging in, and by extending the application to support multiple users, we need to make sure that all the public-facing pages are secured as well. We have already secured the TodosController.index function by calling the authenticated! function, but there are other pages that are not yet secured. We need to make sure that we protect pages accessible over both GET and POST – even if a page is not directly linked to from our app, a malicious user can see what other URLs our app exposes and can try to access them directly to expose and corrupt our data.

As such, besides the TodosController.index function, we need to secure the following functions in the TodosController, by adding the authenticated! function call at the top of each function, exactly like we did for the index function. For instance, for the TodosController. create function, the updated code will look like this:

```
function create()
    authenticated!()          # <----- we have added this line,
                              # the rest is unchanged

    todo = Todo(todo = params(:todo),
                user_id = current_user_id())

    # rest of the function is unchanged
end
```

Apply the same logic to the following functions: TodosController.toggle, TodosController. delete, and TodosController.update.

Securing the API

This should take care of the application's public-facing pages. However, we also need to make sure that the API is secured. Remember that a chain is as strong as its weakest link, and if we don't secure the API, we are leaving a door open for malicious users.

We can just go ahead and apply the same logic as we did for the public-facing pages by adding the `authenticated!` function call to all our public-facing API functions. This would work, but it's not ideal. The problem is that by default, the `authenticated!` function is optimized to support integration with web pages. What this means is that it will redirect the user to the login page if they are not authenticated. This is not what we want for the API, as we want to return a JSON response instead (a redirect response is not valid JSON).

Exactly for this use case, `GenieAuthentication` allows us to specify a custom response to be returned when the user is not authenticated. All we need to do is to create a new JSON response and send it to the unauthenticated users. Add the following code to the `TodosController.API.V1` module, inside the `TodosController.jl` file, for example, right after the `using` statements:

```
using Genie.Exceptions
const NOT_AUTHORISED_ERROR = ExceptionalResponse(401,
["Content-Type" => "application/json"], "Not authorised")
```

With this code, we have created a new `ExceptionalResponse` object, which is a type of error response. We will use it to return a `401` unauthorized response to unauthenticated users, by passing it to the `authenticated!` function. For instance, to secure the `TodosController.API.V1.list` function, we will update the function body to look like this:

```
function list()
  authenticated!(NOT_AUTHORISED_ERROR)

  TodosController.todos() |> json
end
```

That's all there is to it! When attempting to access our API endpoint, unauthenticated users will now receive a `401` response, with a JSON body containing the message `Not authorised`.

Repeat the same logic for all the publicly accessible API functions, by adding the `authenticated!(NOT_AUTHORISED_ERROR)` function call at the top of each of the following functions:

- `TodosController.API.V1.item`
- `TodosController.API.V1.persist`
- `TodosController.API.V1.create`
- `TodosController.API.V1.update`
- `TodosController.API.V1.delete`

Updating our tests

Excellent, our application is now protected against unauthorized access. However, we need to make sure that our tests are updated as well. If we run our test suite now, we will see that all the tests are failing because they are also unable to access authentication-protected pages and API endpoints. So, we need to allow our tests to authenticate as well.

Enabling basic authentication

How can we authenticate our tests – and non-human users in general? We want to allow automation and scripts, like our tests, to access specific user data without having to go through the login process. There are a few methods for authenticating non-human users, and one of the most common and simplest is to use the so-called **basic authentication**. This is a standard HTTP authentication method, which allows us to restrict access to our server using the HTTP Basic schema.

To put it very simply, the Basic schema allows us to send a username and password in the HTTP request header. The header has a standard format: the key is Authorization, and the value is Basic <base64 encoded username:password>. That is, under the Authorization key, we send the Basic schema, followed by a space, followed by the base64 encoded username and password, separated by a colon. (You can read more about the HTTP Basic schema at https://developer.mozilla.org/en-US/docs/Web/HTTP/Authentication.)

In order to enable basic authentication, we need to integrate the dedicated GenieAuthentication features into our app. Add the following at the bottom of the genie_authentication.jl file:

```
# basic auth
basicauth(req, res, params) = begin
  if GenieAuthentication.isbasicauthrequest(params)
    try
      user = findone(Users.User, username =
              params[:username], password =
              Users.hash_password(params[:password]))
      user === nothing && return req, res, params
      login(user, authenticate(user.id,
        GenieAuthentication.GenieSession.session(params)))
    catch _
    end
  end

  req, res, params
```

```
end
basicauth in Genie.Router.pre_match_hooks || push!(Genie.
Router.pre_match_hooks, basicauth)
```

Genie hooks

The code snippet in the preceding subsection uses a Genie hook to enable basic authentication. Hooks are a powerful feature of Genie that is common to many frameworks. They allow us to run custom code at specific points in the request life cycle. That is, we can register functions that will be automatically invoked by Genie when certain events occur. In this case, we register pre_match_hook with the Genie router. Pre-match hooks are functions that are invoked before Genie matches the request to a route. This hook is triggered very early in the request-response life cycle, and it allows us to modify the request and response objects, as well as the request parameters. Router hooks functions are expected to accept three arguments: the request, the response, and the request parameters. The function is expected to return the same three arguments, optionally modifying them.

In our code, we have registered the basicauth function as a pre-match hook. The function checks whether the request is a basic authentication request by looking for the Authorization header. If the request is a basic authentication request, the function attempts to find a user with the given username and password. If a user is found, the function logs the user in and returns a modified request, response, and parameters. If no user is found, the function returns the request, response, and parameters without modifying them.

Setting up the authentication flow

Let's configure the Basic authentication header for our tests. In the test/runtests.jl file, add the following code, right above the @testset block at the bottom of the file:

```
using Base64
const DEFAULT_HEADERS = Dict("Authorization" => "Basic
$(base64encode("testuser:testpass"))")
```

As mentioned, we need to encode the username and password, and we are using the base64encode function from the Base64 module. Base64 is a binary-to-text encoding scheme that transforms binary data in an ASCII string format using a radix -64 representation (for more info, see https://en.wikipedia.org/wiki/Base64).

Then, we simply declare a new constant, DEFAULT_HEADERS, which is a dictionary containing the Authorization header's key and value. We're declaring them here as we'll be using them throughout all of our tests.

Remember to also add `Base64` as a dependency to our tests project. To do this, start a Julia session in the `test/` folder, and run the following command:

```
julia> ] # enter the package manager
pkg> activate .
(test) pkg> add Base64
```

Creating the default test user

For the username and password, we're using `testuser` and `testpass`. However, there's a problem: we don't have a user with this username and password. In addition, remember that the test database is set up before each test run, and is reset after each test suite to ensure that no preexisting state affects the test, so we need to create the test user dynamically before each test run. Given that we already use the migrations to create the database tables, we can use the migrations to create the test user as well. Let's add a new migration. In the `Julia/Genie` app REPL in the root folder, run the following command:

```
julia> using SearchLight
julia> SearchLight.Migration.new("create default user")
```

This will create a new migration file in the `migrations` folder. Open the file and add the following code:

```
module CreateDefaultUser

using Genie
using SearchLight
using ..Main.TodoMVC.Users

function up()
  Genie.Configuration.istest() || return

  Users.User( username = "testuser",
              password = "testpass" |>
              Users.hash_password,
              name     = "Test user",
              email    = "testuser@test.test") |> save!
end

function down()
  Genie.Configuration.istest() || return
```

```
    findone(Users.User, username = "testuser") |> delete
end


end
```

In the up function, we create a new user with the username and password we want to use for our tests. We then save the user to the database. In the down function, we delete the user from the database. Notice that, since we only want to run these functions when we are in test mode, we check for the current environment using the `Genie.Configuration.istest()` function. This function returns `true` if the current environment is `test`, and `false` otherwise. If we're not in test mode, we simply return from the functions without running the actual migration code.

This migration, however, introduces an interesting problem. It needs to have access to application logic, like our `User` model. This means that it can only be run after the application is fully loaded. However, remember that in our app, we run the migrations in the `searchlight.jl` initializer. Initializers are run before the app resources are loaded, including the models. As such, this migration will crash our app upon startup. To address this, again, we will resort to Genie hooks. We need to replace the following snippet in the `searchlight.jl` initializer:

```
try
    SearchLight.Migration.init()
catch
end
SearchLight.Migration.allup()
```

We need to replace that with the following:

```
push!(Genie.Loader.post_load_hooks, () -> begin
    try
        SearchLight.Migration.init()
    catch
    end
    SearchLight.Migration.allup(context = @__MODULE__)
end)
```

Here, instead of directly running the migrations, we wrap the logic into an anonymous function and register it as a `Loader` post-load hook. This hook is invoked by Genie after the app resources are loaded, and it allows us to run code that depends on said resources. We simply delegate the migration logic to the hook function to be executed by Genie at exactly the right time.

Finally, notice that we also pass a `context` value to the `Migration.allup` function, allowing us to inject the dependencies used by the migration.

Wrapping up the tests

Now that we have configured the `Basic` authentication for our app and API, and we have created the test user, we can finally update our tests. Updating the tests needs to cover the following areas:

- Set up and tear down the test database before and after each test suite, passing in the `context` value to inject the app's dependencies

- Add the `Authorization` header to all the requests, so that the requests are authenticated

- Alter all the creation/persistence code for to-do items to make sure that the `user_id` value is set to the ID of the test user

Let's proceed. Starting with the `test/todos_test.jl` file, modify the existing `@testset "Todo is correctly initialized"` and `@testset "Todo is valid"` blocks:

```julia
@testset "Todo is correctly initialized" begin
    @test t.todo == ""
    @test t.completed == false
    @test t.user_id.value == nothing # <-------------------
                                     # add this line
end
@testset "Todo is valid" begin
    t.todo = "Buy milk"
    t.user_id = 1  # <-------------------- add this line
    v = validate(t)

    # rest of the code the same
end
```

As for the `test/todos_API_test.jl`, `test/todos_db_test.jl`, and `test/todos_integration_test.jl` files, we'll refer to the code in the application repository (see the link in the *Technical requirements* section at the start of this chapter).

As before, run the tests at the command line with the following:

```
$ julia --project runtests.jl
```

You'll see an output like this:

```
todos_API_test: ............................
todos_db_test: .....
todos_integration_test: WARNING: redefinition
 produce other errors.
.......
todos_test: ........

Test Summary: | Pass  Total   Time
TodoMVC tests |   99     99  45.2s
```

Figure 7.4 – Results of the test run

Summary

Access to our app and data needs to be secure. In the Genie framework, this security is provided by the `GenieAuthentication` plugin.

In this chapter, we learned how to configure this plugin and set up the database accordingly. Then, we coded functionality for registering a user, so that we could restrict access to the app itself. The next step was restricting access to the app's data itself. Finally, we updated our tests to prove that everything still works.

In the next chapter, we'll finally add the eagerly awaited dashboard functionality to our ToDo app!

8

Developing Interactive Data Dashboards with Genie

Julia is a relatively new programming language, but it has already gained a lot of traction in the data science community. It is designed to be easy to learn and use and has seen great adoption in digital R&D, scientific computing, data analysis, and machine learning. A critical part of the data science workflow is the ability to create interactive data dashboards for data exploration and analysis.

In this chapter, we will see how to use the Genie framework to create interactive data dashboards. By the end of this chapter, you'll be able to start building data dashboards for your own projects.

In this chapter, we will discuss the following topics:

- Augmenting data
- Generating random data
- Building our data dashboard
- Adding the final touches

Technical requirements

The source code for this chapter can be found at `https://github.com/PacktPublishing/Web-Development-with-Julia-and-Genie/tree/main/Chapter8`. Please make sure you use *Genie v5.2* or higher.

We also need the `GenieFramework` plugin in order to make a dashboard, so open the Julia REPL in the app's root folder, type] to go into Pkg mode, and enter the following:

```
(@v1.8) pkg> add GenieFramework
(@v1.8) pkg> update
```

Augmenting data

We will extend our to-do app with a new section that will allow us to analyze and visualize our to-do list to understand how our time is allocated between various types of activities. However, as it is right now, our data is not very useful for this purpose. We need richer data, and a lot more of it, in order to do interesting things. So, let's add a few more fields to our to-do items and generate a lot of random data.

To ensure an interesting analysis, let's add the following fields:

- category: We'll make this into a string with one of the following values: "work", "personal", "family", "hobby", "errands", "shopping", "accounting", "learning", or "other"
- date: The day the to-do item was created
- duration: An integer representing the duration of the to-do item in minutes

And for our data dashboard, we will allow our users to filter the data by category and date interval, and visualize the to-do data by date, category, and duration, exposing interesting stats about individual to-dos (such as due, overdue, and so on) and aggregated stats (such as categories that take the most time or the total time by day).

We'll begin by adding the new fields to our to-do items. As usual, we'll use a migration, so let's create it:

```
julia> using SearchLight
julia> SearchLight.Migrations.new("add category date and
duration to todos")
```

Once the migration file is created in /db/migrations/datetime_add_category_date_ and_duration_to_todos.jl, edit the file as follows:

```
module AddCategoryDateAndDurationToTodos

import SearchLight.Migrations: add_columns, remove_columns,
add_indices, remove_indices

function up()
  add_columns(:todos, [
    :category => :string,
    :date => :date,
    :duration => :int
  ])
```

```
  add_indices(:todos, [
    :category,
    :date,
    :duration
  ])
end

function down()
  remove_indices(:todos, [
    :category,
    :date,
    :duration
  ])

  remove_columns(:todos, [
    :category,
    :date,
    :duration
  ])
end

end
```

When ready, run the migration to update the database schema:

```
julia> SearchLight.Migrations.up()
```

Now that we have the new fields in our database, we need to update our to-do model to reflect the new fields. Update the app/resources/todos/Todos.jl file by replacing the declaration of the Todo struct with the following:

```
using Dates

const CATEGORIES = ["work", "personal", "family", "hobby",
"errands", "shopping", "accounting", "learning", "other"]

@kwdef mutable struct Todo <: AbstractModel
  id::DbId = DbId()
```

```
    todo::String = ""
    completed::Bool = false
    user_id::DbId = DbId()
    category::String = CATEGORIES[end]
    date::Date = Dates.today()
    duration::Int = 30
end
```

At the start, we added a new dependency in the `Dates` module, as our new `date` property is a `date` instance. This means that we also need to declare the dependency in our `Project.toml` file, so make sure to run the following in the app's REPL:

```
pkg> add Dates
```

It is also necessary to add `using Dates` at the top of the module, near the other `using` and `import` statements. We also defined a `CATEGORIES` constant, where we stored the list of possible categories. We then updated the `Todo` struct to include the new fields, setting their default values to the last category in the list (`"other"`), today's date, and `30` minutes.

Generating random data

Now that our database and model definition have been updated, we need to generate some random data to populate our database and make our dashboard more interesting. We also need to set some random values for the new columns of the to-dos that already exist in the database. Let's create a new migration to script and run our data generation:

```
julia> using SearchLight
julia> SearchLight.Migrations.new("generate fake todos")
```

In the resulting migration file, add the following code:

```
module GenerateFakeTodos

using ..Main.TodoMVC.Todos
using SearchLight
using Dates
using Faker

randcategory() = rand(Todos.CATEGORIES)
```

```
randdate() = Dates.today() - Day(rand(0:90))
randduration() = rand(10:240)

function up()
  for i in 1:1_000
    Todo(
      todo = Faker.sentence(),
      completed = rand([true, false]),
      user_id = DbId(1),
      category = randcategory(),
      date = randdate(),
      duration = randduration()
    ) |> save!
  end

  for t in find(Todo, SQLWhereExpression("category is ?",
                nothing))
    t.category = randcategory()
    date = randdate()
    duration = randduration()
    save!(t)
  end
end

function down()
  throw(SearchLight.Migration.
IrreversibleMigrationException(@__MODULE__))
end

end
```

This migration will generate 1,000 new to-dos with random values and then update the existing to-dos with more random values for the new columns. We use the Faker package to generate random sentences for the to-do field, so make sure to add it to the app file as well:

```
pkg> add Faker
```

For the generation of the random categories, dates, and duration values, we declare three helper functions (`randcategory`, `randdate`, and `randduration`) that we can reuse to both create new to-dos and update the existing ones.

Now (after closing and reopening the Julia REPL), you can run the migration to generate the fake data:

```
julia> using SearchLight
julia> SearchLight.Migrations.up()
```

Building our data dashboard

Now that we have a lot more data, we can start building our data dashboard. We'll start by creating a new resource (controller) for our dashboard:

```
julia> using Genie
julia> Genie.Generator.newresource("dashboard", pluralize =
false)
```

This will create a new controller file, `app/resources/dashboard/DashboardController.jl`.

Using Genie with low-code and reactive programming

For the `dashboard` component, we will introduce a new paradigm of programming, by using low code to create reactive web user interfaces. What does this mean? When creating our to-do list, we used pure web development techniques, writing low-level web user interfaces using HTML, CSS, and JavaScript. However, Genie provides a more productive way of creating web user interfaces using low-code and reactive programming. This way, we can create web UIs without having to write HTML and JavaScript code. We do this by employing a series of Genie packages that give us access to around 100 UI components, covering all necessary elements of a web application, including inputs and forms (dropdowns, sliders, ranges, and so on), text and content, and plots. These components are reactive, meaning that they automatically update based on user input, without needing to do a page refresh. This is ideal for data exploration dashboards, where we want to be able to interact with the data and see the results immediately.

These features are provided by a package called `GenieFramework`. Genie is a part of the **GenieFramework** set of libraries, but it also exposes the aforementioned extra packages for low-code and web UI components and reactive programming.

So, let's start working on our dashboard. First, we need to add the `GenieFramework` package to our app's `Project.toml` file:

```
pkg> add GenieFramework
```

Now, we can edit our `DashboardController.jl` file:

```
module DashboardController

using GenieFramework
using TodoMVC, TodoMVC.Todos
using Dates
using GenieAuthentication


end
```

In the preceding code, we have declared the dependencies. Besides the `GenieFramework` package, we also need to import our `Todo` model, the `Dates` module, and the `GenieAuthentication` package, which we will use to protect our dashboard with authentication.

Next, under the last `using` declaration, add the following code:

```
@handlers begin
  @in filter_startdate = today() - Month(1)
  @in filter_enddate = today()

  @out total_completed = 0
  @out total_incompleted = 0
  @out total_time_completed = 0
  @out total_time_incompleted = 0

  @out todos_by_status_number = PlotData[]
  @out todos_by_status_time = PlotData[]
  @out todos_by_category_complete = PlotData[]
  @out todos_by_category_incomplete = PlotData[]
end
```

Central to the low-code and reactive programming paradigm is the use of `@handlers` and `@in` and `@out` declarations. The `@handlers` block designates the code that will be executed when the dashboard is loaded and when the user interacts with the various inputs on the page. The `@in` and `@out` declarations define the input and output variables that will be used in the dashboard. The `@in` variables are the ones that will be used to filter the data, and the `@out` variables will be used to display the results.

So, we start by declaring the data – what inputs we want to receive and what outputs we want to display. Later on, we'll add reactive handlers to some of these values to update the data and output the results again when the inputs change.

We said that we want to allow the users to filter the to-dos by the date range, so we declare two input variables, `filter_startdate` and `filter_enddate`. These are inputs, so we use the `@in` declaration. We also declare four output variables, `total_completed`, `total_incompleted`, `total_time_completed`, and `total_time_incompleted`, which will be used to display their corresponding data as **Big Number components** on our dashboard. *Big Number* is a common term used in dashboard terminology to indicate visual components that only show numbers to draw the attention of managers to key metrics in a report. These values should be pretty much self-explanatory, but just in case it's not entirely clear, `total_completed` will represent the total number of completed to-dos, `total_incompleted` will represent the total number of incomplete to-dos, and `total_time_completed` and `total_time_incompleted` will represent the total time spent on completed and incomplete to-dos, respectively.

In addition, we'll also want to display a number of plots (charts), as follows:

- An area chart with two overlapping areas showing the number of completed and incomplete to-dos over time
- A stacked bar chart showing the number of completed and incomplete to-dos and their duration, also over time
- Two pie charts showing the distribution of completed and incomplete to-dos by category

The plots use `PlotData` vectors to display their data, so we declare four `@out` variables, `todos_by_status_number`, `todos_by_status_time`, `todos_by_category_complete`, and `todos_by_category_incomplete`, one for each of the charts we want to display.

Each of the 10 reactive variables we declared will be automatically synchronized between the server and the browser automatically, every time we update them. The `@in` values accept changes from the browser (but we can also update them from the server and they will be reflected in the UI), while the `@out` values can't be set from the browser, only from the server.

Finally, note that we initialize all our variables to empty default values. We'll use event handlers later on to update them to their proper values.

However, now that we have set up our reactive variables, we can design our view using Genie's low-code API. Because our values are set to empty default values, we won't see the actual data, but it's enough to allow us to build our UI and make sure it works.

Designing the dashboard view

We'll start by adding a new view file, app/resources/dashboard/views/index.jl:

```
container([

  # section 1 #
  btn(color="primary", flat=true, "⇐ Home",
      onclick="javascript:window.location.href='/';")
  h1("Todos productivity report")
  # end section 1 #

  # section 2 #
  # date filters row
  row([
    cell(class="col-6", [
      textfield("Start date", :filter_startdate,
                clearable = true, filled = true, [
        icon(name = "event", class = "cursor-pointer",
            style = "height: 100%;", [
          popup_proxy(cover = true,
                      [datepicker(:filter_startdate,
                      mask = "YYYY-MM-DD")])
        ])
      ])
    ])

    cell(class="col-6", [
      textfield("End date", :filter_enddate,
                clearable = true, filled = true, [
                icon(name = "event",
                class = "cursor-pointer",
                style = "height: 100%", [
                popup_proxy(ref = "qDateProxy",
                cover = true, [datepicker(:filter_enddate,
                              mask = "YYYY-MM-DD")])
      ])
```

```
        ])
      ])
    ])
    # end date filters row
    # end section 2 #

    # section 3 #
    # big numbers row
    row([
      cell(class="st-module", [
        row([
          cell(class="st-br", [
            bignumber("Total completed", :total_completed,
                      icon="format_list_numbered",
                      color="positive")
          ])
          cell(class="st-br", [
            bignumber("Total incomplete", :total_incompleted,
                      icon="format_list_numbered",
                      color="negative")
          ])
          cell(class="st-br", [
            bignumber("Total time completed",
                      :total_time_completed,
                      icon="format_list_numbered",
                      color="positive")
          ])
          cell(class="st-br", [
            bignumber("Total time incomplete",
                      :total_time_incompleted,
                      icon="format_list_numbered",
                      color="negative")
          ])
        ])
      ])
    ])
```

```
# end big numbers row
# end section 3 #

# section 4 #
row([ # area chart -- number of todos by status
  cell(class="st-module col-12", [
    plot(:todos_by_status_number,
          layout = "{ title: 'Todos by status',
                    xaxis: { title: 'Date' }, yaxis: {
                    title: 'Number of todos' } }")
  ])
]) # end area chart -- number of todos by status

row([ # stacked bar chart -- duration of todos by status
  cell(class="st-module col-12", [
    plot(:todos_by_status_time, layout = "{ barmode:
          'stack', title: 'Todos by status and duration',
          xaxis: { title: 'Date' },
          yaxis: { title: 'Total duration' } }")
  ])
])  # end stacked bar chart -- duration of todos by
    # status

row([
  # pie chart -- number of completed todos by category
  cell(class="st-module col-6", [
    plot(:todos_by_category_complete, layout = "{ title:
          'Completed todos by category', xaxis: { title:
          'Category' }, yaxis: { title: 'Number of todos'
          } }")
  ])

  # pie chart -- number of incomplete todos by category
  cell(class="st-module col-6", [
    plot(:todos_by_category_incomplete, layout = "{
          title: 'Incompleted todos by category',
```

```
            xaxis: { title: 'Category' },
            yaxis: { title: 'Number of todos' } }")
    ])
  ])
  # end section 4 #
])
```

Let's break down the preceding code. First, everything is wrapped in a `container`. Containers are used for building responsive layouts, in conjunction with rows and columns. Rows are used to create horizontal groups of columns, and columns are used to create vertical groups of content. We'll use rows and columns to create a responsive layout for our dashboard.

Then, in the first section, we add a button to go back to the home page, and a title. The second section contains the date filters. Section 3 contains the big numbers row, and section 4 contains the charts.

For the date filters, we use a text input field with a date picker. We use the `popup_proxy` component to create a popup that will display the date picker when the user clicks on the calendar icon. Also, we bind the value of the text field and the date picker to the same reactive variable, `:filter_startdate`, so that whenever the user changes the date in the date picker or the text field, our `filter_startdate` reactive variable on the server will be updated.

Big numbers are very commonly used UI elements for data dashboards. As the name suggests, they display important values together with a label and an icon. We bind them to their corresponding server-side values.

Then, in the last section, we have the plots. We display them across three rows, one for the area chart, one for the stacked bar chart, and one for the two pie charts. All we have to do in order to display the plots is to bind them to their corresponding reactive variables. We also customize the layout of the plots, passing in various extra options.

That's all – our UI is ready! Note how simple it is to create powerful UIs with Genie's low-code API, using a declarative syntax in pure Julia. We didn't have to write a single line of HTML or JavaScript!

Updating the dashboard view

Now that we have our UI ready, we need to update it with the actual data. Back in our `DashboardController.jl` file, inside the `@handlers` block, under the last reactive variable declaration, add the following code:

```
@onchangeany isready, filter_startdate, filter_enddate begin

# we'll add more code here
```

```
end
```

With the preceding code, we have declared an event handler that will be triggered whenever any of the reactive variables, isready, filter_startdate, or filter_enddate, change. The filter_startdate and filter_enddate reactive variables are the ones we connected to the date filters in the UI, while the isready reactive variable is a Boolean value that is automatically provided and updated by Genie itself. By default, it's set to false, and when the UI is fully rendered, loaded, and ready to receive data from the server, it's automatically set to true. So what this event handler means is that whenever the dashboard is loaded or the user changes the date filters, we update the dashboard view with fresh data.

Inside the event handler, add the following code:

```
completed_todos = Todos.search(; completed = true, startdate =
filter_startdate, enddate = filter_enddate)

incompleted_todos = Todos.search(; completed = false, startdate
= filter_startdate, enddate = filter_enddate)

completed_todos_by_category = Todos.search(; completed = true,
group = ["category"], startdate = filter_startdate, enddate =
filter_enddate)

incompleted_todos_by_category = Todos.search(; completed =
false, group = ["category"], startdate = filter_startdate,
enddate = filter_enddate)


# more code will come here
```

We start by computing the completed and incomplete to-dos, and the completed and incomplete to-dos by category. We will need these values to pull out the various stats and populate our big numbers and our plots. We use the Todos.search function to do that, passing in the completed and group parameters to filter the to-dos by status and category. We also pass in the startdate and enddate parameters to filter the to-dos by date. The Todos.search function returns a DataFrame object, which is a tabular data structure that is very similar to a spreadsheet.

Let's create the Todos.search function. Open the Todos.jl model file and add the following code at the end of the module (inside the module, after the validators block):

```
function search(; completed = false, startdate = today() -
Month(1), enddate = today(), group = ["date"], user_id)
```

```
    filters = SQLWhereEntity[
        SQLWhereExpression("completed = ?", completed),
        SQLWhereExpression("date >= ? AND date <= ?",
                            startdate, enddate),
    SQLWhereExpression("user_id = ?", user_id)
    ]


    DataFrame(Todo, SQLQuery(
      columns = SQLColumns(Todo, (
        total_time = SQLColumn("SUM(duration) AS
                     total_time", raw = true),
        total_todos = SQLColumn("COUNT(*) AS
                     total_todos", raw = true),
      )),
      where = filters,
      group = group,
      order = ["date ASC", "category ASC"],
    ))
  end
```

In addition, add the following at the top of the module (besides the other using and import statements):

```
julia> import SearchLight.DataFrames.DataFrame
```

Now, going back to the DashboardController.jl file, let's add more code to our event handler:

```
total_completed = sum(completed_todos[!,:total_todos])
total_incompleted = sum(incompleted_todos[!,:total_todos])
total_time_completed = sum(completed_todos[!,:total_time]) / 60
|> round
total_time_incompleted = sum(incompleted_todos[!,:total_time])
/ 60 |> round
```

We compute the total number of completed and incomplete todos, and the total time spent on completed and incomplete to-dos. We use the sum function to sum up the values in the total_todos and total_time columns of the completed_todos and incompleted_todos DataFrame.

Next, let's add more code to our event handler:

```
todos_by_status_number = [
  PlotData(
    x = completed_todos[!,:todos_date],
    y = completed_todos[!,:total_todos],
    fill = "tozeroy",
    name = "Completed",
    plot = StipplePlotly.Charts.PLOT_TYPE_SCATTER
  ),

  PlotData(
    x = incompleted_todos[!,:todos_date],
    y = incompleted_todos[!,:total_todos],
    fill = "tozeroy",
    name = "Incompleted",
    plot = StipplePlotly.Charts.PLOT_TYPE_SCATTER
  ),
]
```

This is the data for the first plot, the area chart that shows the number of completed and incomplete to-dos over time.

Moving on, let's create the rest of our chart's data:

```
todos_by_status_time = [
  PlotData(
    x = completed_todos[!,:todos_date],
    y = completed_todos[!,:todos_duration],
    name = "Completed",
    plot = StipplePlotly.Charts.PLOT_TYPE_BAR
  ),

  PlotData(
    x = incompleted_todos[!,:todos_date],
    y = incompleted_todos[!,:todos_duration],
    name = "Incompleted",
    plot = StipplePlotly.Charts.PLOT_TYPE_BAR
```

```
  ),
]

todos_by_category_complete = [
  PlotData(
    values = completed_todos_by_category[!,:total_todos],
    labels =
      completed_todos_by_category[!,:todos_category],
    plot = StipplePlotly.Charts.PLOT_TYPE_PIE
  )
]

todos_by_category_incomplete = [
  PlotData(
    values = incompleted_todos_by_category[!,:total_todos],
    labels =
      incompleted_todos_by_category[!,:todos_category],
    plot = StipplePlotly.Charts.PLOT_TYPE_PIE
  )
]
```

Adding the routes

Now that our controller logic is ready and our UI is built, let's set up the route to render the dashboard view. Go to the routes.jl file and add the following at the bottom:

```
route("/dashboard", DashboardController.index)
```

We indicate that the /dashboard route should render the DashboardController.index function. So, let's create the index function at the bottom of the DashboardController.jl file (under the @handlers block, inside the module):

```
function index()
  authenticated!()

  @page("/dashboard",
  "app/resources/dashboard/views/index.jl").route.action()
end
```

We want to make sure that access is restricted to authenticated users, so we call the authenticated! function. Then, we build a page using the @page macro (also part of Genie's low-code API). You can think of the page as a mix of routes and views. It abstracts away a lot of complexity by setting up all the necessary pieces to enable the reactive UI. We create a new instance of a page, indicating that the route is /dashboard and the view is app/resources/dashboard/views/index.jl.

We then call the action function on the page's route to render the view – that's all.

First, run the following command:

```
pkg> instantiate
```

Our dashboard is now up and running, as you can see in the following screenshots.

The first screenshot shows the overview screen, showing all to-do data and functionality:

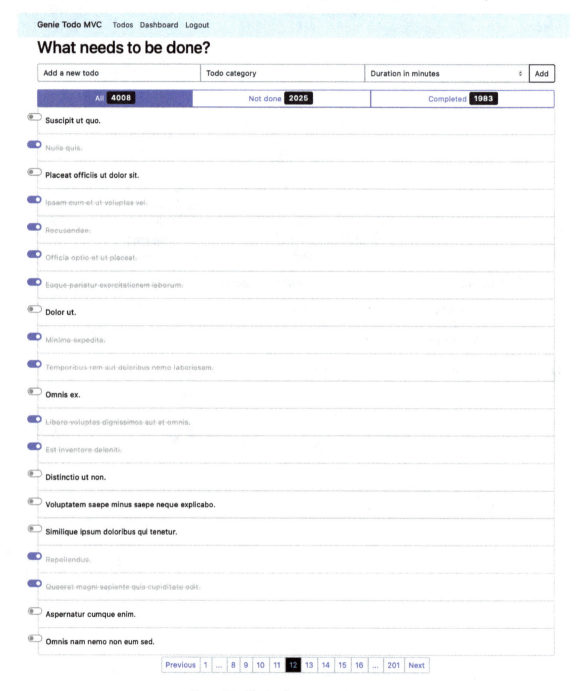

Figure 8.1 – The to-do app overview screen

Then, by clicking the **Dashboard** link in the top menu, you get the following data-analysis view:

Figure 8.2 – The Dashboard screen

Adding the final touches

Our to-do app is complete. However, we can make it even better by adding a few more features.

Adding pagination to the to-do list

If we go back to the to-do list, we can see that now we have a lot of data. It's getting hard to navigate through the list. Let's add pagination to make the data more manageable.

First, let's compute the pagination data, calculating the total number of pages and the logic for the previous and next buttons. In the `TodosController.jl` file, add the following:

```
const TODOS_PER_PAGE = 20
const PAGINATION_DISPLAY_INTERVAL = 5
const MAX_PAGINATION_WIDTH = 30


page() = parse(Int, params(:page, "1"))


function count_pages()
  total_pages = count(Todo, user_id = current_user_id()) /
                     TODOS_PER_PAGE |> ceil |> Int
  current_page = page()
  prev_page = current_page - 1
  next_page = current_page < total_pages ? current_page + 1 : 0


  (
    total_pages = total_pages,
    current_page = current_page,
    prev_page = prev_page,
    next_page = next_page
  )
end
```

Then, in the `todos()` function, we need to update the logic to take into account the pagination:

```
function todos()
  todos = if params(:filter, "") == "done"
    find(Todo, completed = true, user_id =
        current_user_id())
```

```
  elseif params(:filter, "") == "notdone"
    find(Todo, completed = false, user_id =
        current_user_id())
  else
    # this is updated to take into account the pagination
    find(Todo;  limit = TODOS_PER_PAGE |> SQLLimit,
                offset = (page() - 1) * TODOS_PER_PAGE,
                user_id = current_user_id(),
                order = "date DESC")
  end
end
```

Next is the pagination UI. Create a new view partial in `app/resources/todos/views/_`
`pagination.jl.html` and add the following code:

```
<nav>
  <ul class="pagination pagination-sm"
    style="justify-content: center;">
    <% iif(prev_page > 0) do %>
      <li class="page-item"><a class="page-link"
        href="/?page=$prev_page">Previous</a></li>
    <% end %>
    <% iif(total_pages > MAX_PAGINATION_WIDTH &&
      current_page > PAGINATION_DISPLAY_INTERVAL) do ;[ %>
      <li class="page-item"><a class="page-link"
        href="/?page=1">1</a></li>
      <li class="page-item"><a class="page-link">...</a>
      </li>
    <% ]; end %>
    <% for_each(1:total_pages) do page %>
      <% iif(page >
        (current_page - PAGINATION_DISPLAY_INTERVAL) && page
        < (current_page + PAGINATION_DISPLAY_INTERVAL)) do
        %>
        <li class="page-item">
          <% if page == current_page %>
            <a class="page-link bg-dark
```

```
              text-info">$page</a>
          <% else %>
            <a class="page-link"
              href="/?page=$page">$page</a>
          <% end %>
        </li>
      <% end %>
    <% end %>
    <% iif(total_pages > MAX_PAGINATION_WIDTH &&
      current_page < (total_pages -
      PAGINATION_DISPLAY_INTERVAL)) do ;[ %>
      <li class="page-item"><a class="page-link">...</a>
      </li>
      <li class="page-item"><a class="page-link"
        href="/?page=$total_pages">$total_pages</a></li>
    <% ]; end %>
    <% iif(next_page > 0) do %>
      <li class="page-item"><a class="page-link"
        href="/?page=$next_page">Next</a></li>
    <% end %>
  </ul>
</nav>
```

The preceding code is quite dense, as the pagination includes a lot of logic. Let's break it down:

- If we're not on the first page, we show a **Previous** button

- If there are lots of pages, we replace some of the buttons with ..., between the first page and the current page

- Then, we show the buttons around the current page and mark the page with a different color

- Next, if there are lots of pages between the current page and the last page, we replace some of the buttons with ... and show the last page

- Finally, if we're not on the last page, we show a **Next** button

Updating the to-do item creation

Another thing we need to take into account is that we have added extra properties to our to-do items (category, date, and duration). However, we haven't provided a way for our users to set these properties when creating a new to-do item. Let's do it now.

Start by editing the view partial in app/resources/todos/views/_form.jl.html and make it look like this:

```
<div class="row">
  <form method="POST" action="/todos"
    class="form-floating">
    <div class="input-group mb-3">
      <input type="text" class="form-control"
        placeholder="Add a new todo" name="todo"
        value='$(params(:todo, ""))' />
      <input class="form-control" list="categories_list"
        name="category" placeholder="Todo category"
        value='$(params(:category, ""))' />
      <datalist id="categories_list">
        <% for_each(categories) do category %>
          <option value="$category" />
        <% end %>
      </datalist>
      <input type="number" class="form-control"
        placeholder="Duration in minutes" name="duration"
        value='$(params(:duration, ""))' min="5" max="240"
        step="5" />
      <input type="submit"
        class="btn btn-outline-secondary" value="Add" />
    </div>
  </form>
</div>
```

In the preceding code, we have added a number of fields to the form:

- A text field for the category, with a list of existing categories to choose from. The data for datalist is provided by a categories variable that we'll need to pass into the view.

- A number field for the duration, with a minimum of 5 minutes, a maximum of 240 minutes, and a step of 5 minutes.

- A date field for the date, with the current date as the default value.

In TodosController, we need to update the create() function to take into account the new fields:

```
function create()
  authenticated!()

  # we add the new values here
  todo = Todo(todo = params(:todo),
          category = params(:category),
          duration = params(:duration),
          user_id = current_user_id())

  # rest of the code is the same
end
```

We also need to add the logic to get the list of categories used on the frontend – add this function to TodosController:

```
categories() = vcat(SearchLight.query("SELECT DISTINCT category
from todos ORDER by category ASC")[!,:category], Todos.
CATEGORIES) |> unique! |> sort!
```

Finally, the index() function also needs to be updated to pass the pagination and the categories data to the view:

```
function index()
  authenticated!()

  html(:todos, :index; todos = todos(), count_todos()...,
        count_pages()..., ViewHelper.active,
        MAX_PAGINATION_WIDTH,
        PAGINATION_DISPLAY_INTERVAL,
```

```
            categories = categories())
  end
```

Adding navigation

We have a lot of pages now, and it's getting a bit hard to navigate between them. Let's add a navigation bar to the top of the page.

Start by editing the view partial in app/layouts/_main_menu.jl.html and make it look like this:

```
<nav class="navbar navbar-expand-lg navbar-light"
style="background-color: #e3f2fd;">
  <div class="container-fluid">
    <a class="navbar-brand" href="#">Genie Todo MVC</a>
    <button class="navbar-toggler" type="button"
      data-bs-toggle="collapse"
      data-bs-target="#navbarSupportedContent">
      <span class="navbar-toggler-icon"></span>
    </button>
    <div class="collapse navbar-collapse"
      id="navbarSupportedContent">
      <ul class="navbar-nav me-auto mb-2 mb-lg-0">
        <li class="nav-item">
          <a class="nav-link" href="/">Todos</a>
        </li>
        <li class="nav-item">
          <a class="nav-link"
            href="/dashboard">Dashboard</a>
        </li>
        <li class="nav-item">
          <a class="nav-link" href="/logout">Logout</a>
        </li>
      </ul>
    </div>
  </div>
</nav>
```

The preceding code is for a simple navigation bar with three links: **Todos**, **Dashboard**, and **Logout**. We added it to the `layouts/` folder because we want it to be available on all pages. So, let's include the partial in the app's layout. Update the `<body>...</body>` part of the `app/layouts/app.jl.html` file to look like this:

```
... more code here ...
<body>
  <div class="container-fluid">
    <% partial("app/layouts/_main_menu.jl.html") %>
    <div class="container">
      <h1>What needs to be done?</h1>
      <%
        @yield
      %>
    </div>
  </div>
  <script src=
    "https://unpkg.com/axios/dist/axios.min.js"></script>
  <script src="https://cdn.jsdelivr.net/npm/cash-
              dom@8.1.1/dist/cash.min.js">
  </script>
  <script src="/js/app.js"></script>
</body>
... rest of the code here ...
```

Updating the to-do list

Now that we have prepared all the logic and the views for the pagination and the updated to-do item creation, let's update the to-do's view to use them. Edit the `view` file in `app/resources/todos/views/index.jl.html` and make it look like this:

```
<% partial("app/resources/todos/views/_messages.jl.html") %>
<% partial("app/resources/todos/views/_form.jl.html") %>
<% partial("app/resources/todos/views/_filters.jl.html") %>
<% if isempty(todos) && params(:filter, "") == "done" %>
    <p>You haven't completed any todos yet.</p>
<% elseif isempty(todos) %>
    <p>Nothing to do!</p>
```

```
<% else %>
  <div class="row">
    <div class="col">
      <ul class="list-group">
        <% for_each(todos) do todo %>
          <li class="list-group-item list-group-item-action
                    form-check form-switch">
            <input type="checkbox"
              checked="$(todo.completed ? true : false)"
              class="form-check-input" id="todo_$(todo.id)"
              value="$(todo.id)" />
            <label class='form-check-label
              $(todo.completed ? "completed" : "")'
              data-original="$(todo.todo)"
              data-todo-id="$(todo.id)">$(todo.todo)
            </label>
            <button
              class="btn btn-outline-danger invisible"
              type="button"
              value="$(todo.id)">Delete</button>
          </li>
        <% end %>
      </ul>
    </div>
  </div>
<% end %>
<% partial("app/resources/todos/views/_pagination.jl.html") %>
```

There is nothing complicated in the preceding code. We have only moved the _filters.jl.html partial to the top of the page so that the user can filter the to-dos without needing to scroll to the bottom. Furthermore, the bottom of the page is now used by the _pagination.jl.html partial to output the pagination links.

Redirecting after login

The final tweak we'll make is to redirect the user to the to-do list after they log in. Currently, we are using the default implementation from `GenieAuthentication`, which redirects to the success page. However, this isn't useful for us, so we're better off going straight to the to-do list after login. Update the `AuthenticationController.jl` file, replacing one line in the `login()` function:

```
function login()
  try
    user = findone(User, username = params(:username),
      password = Users.hash_password(params(:password)))
    authenticate(user.id, GenieSession.session(params()))

    # redirect(:success)  <-- replace this
    redirect("/")         # <-- with this
  catch ex
    flash("Authentication failed! ")

    redirect(:show_login)
  end
end
```

Summary

In this chapter, we used the `GenieFramework` tool to build a dashboard To-Do application. First, we added some new to-do fields, which lend themselves to filtering and plotting. Then, we learned how to generate random data through a dedicated migration.

We applied the reactive web UI techniques from the low-code GenieFramework to build our dashboard view, complete with filters, bar charts, and pie charts.

We finished up by updating our views for the new fields, and by adding pagination, navigation, and redirection after login.

This concludes our overall survey of Julia for the web and the Genie web framework and technologies. We hope you've enjoyed this trip through Julia web-land as much as we did, and urge you to try it out for your next project!

Index

`Packt.com`

Subscribe to our online digital library for full access to over 7,000 books and videos, as well as industry leading tools to help you plan your personal development and advance your career. For more information, please visit our website.

Why subscribe?

- Spend less time learning and more time coding with practical eBooks and Videos from over 4,000 industry professionals

- Improve your learning with Skill Plans built especially for you

- Get a free eBook or video every month

- Fully searchable for easy access to vital information

- Copy and paste, print, and bookmark content

Did you know that Packt offers eBook versions of every book published, with PDF and ePub files available? You can upgrade to the eBook version at `packt.com` and as a print book customer, you are entitled to a discount on the eBook copy. Get in touch with us at `customercare@packtpub.com` for more details.

At `www.packt.com`, you can also read a collection of free technical articles, sign up for a range of free newsletters, and receive exclusive discounts and offers on Packt books and eBooks.

Other Books You May Enjoy

If you enjoyed this book, you may be interested in these other books by Packt:

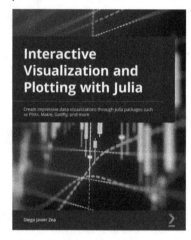

Interactive Visualization and Plotting with Julia

Diego Javier Zea

ISBN: 978-1-80181-051-7

- Create interactive plots with Makie, Plots, Jupyter, and Pluto
- Create standard statistical plots and visualize clustering results
- Plot geographically distributed and biological data
- Visualize graphs and networks using GraphRecipes and GraphPlots
- Find out how to draw and animate objects with Javis, Plots, and Makie
- Define plot themes to reuse plot visual aspect customizations
- Arrange plots using Plots, Makie, and Gadfly layout systems
- Define new plot types and determine how Plots and Makie show objects

Rust Web Programming

Maxwell Flitton

ISBN: 978-1-80056-081-9

- Structure scalable web apps in Rust in Rocket, Actix Web, and Warp
- Apply data persistence for your web apps using PostgreSQL
- Build login, JWT, and config modules for your web apps
- Serve HTML, CSS, and JavaScript from the Actix Web server
- Build unit tests and functional API tests in Postman and Newman
- Deploy the Rust app with NGINX and Docker onto an AWS EC2 instance

Packt is searching for authors like you

If you're interested in becoming an author for Packt, please visit authors.packtpub.com and apply today. We have worked with thousands of developers and tech professionals, just like you, to help them share their insight with the global tech community. You can make a general application, apply for a specific hot topic that we are recruiting an author for, or submit your own idea.

Share your thoughts

Once you've read *Web Development with Julia and Genie*, we'd love to hear your thoughts! Scan the QR code below to go straight to the Amazon review page for this book and share your feedback.

https://packt.link/r/1-801-81113-X

Your review is important to us and the tech community and will help us make sure we're delivering excellent quality content.

Download a free PDF copy of this book

Thanks for purchasing this book!

Do you like to read on the go but are unable to carry your print books everywhere?

Is your eBook purchase not compatible with the device of your choice?

Don't worry, now with every Packt book you get a DRM-free PDF version of that book at no cost.

Read anywhere, any place, on any device. Search, copy, and paste code from your favorite technical books directly into your application.

The perks don't stop there, you can get exclusive access to discounts, newsletters, and great free content in your inbox daily!

Follow these simple steps to get the benefits:

1. Scan the QR code or visit the link below:

https://packt.link/free-ebook/9781801811132

2. Submit your proof of purchase
3. That's it! We'll send your free PDF and other benefits to your email directly

www.ingramcontent.com/pod-product-compliance
Lightning Source LLC
Chambersburg PA
CBHW060539060326
40690CB00017B/3545